HYDRAULIC ENGINEERING IV

PROCEEDINGS OF THE 4TH INTERNATIONAL TECHNICAL CONFERENCE ON HYDRAULIC ENGINEERING (CHE 2016, HONG KONG, 16–17 JULY 2016)

Hydraulic Engineering IV

Editor

Liquan Xie

Department of Hydraulic Engineering, Tongji University, Shanghai, China

CRC Press
Taylor & Francis Group
Boca Raton London New York Leiden

CRC Press is an imprint of the
Taylor & Francis Group, an **informa** business

A BALKEMA BOOK

CRC Press/Balkema is an imprint of the Taylor & Francis Group, an informa business

© 2016 Taylor & Francis Group, London, UK

Typeset by V Publishing Solutions Pvt Ltd., Chennai, India

Published by: CRC Press/Balkema
P.O. Box 11320, 2301 EH Leiden, The Netherlands
e-mail: Pub.NL@taylorandfrancis.com
www.crcpress.com – www.taylorandfrancis.com

ISBN: 978-1-138-02948-4 (Hbk)
ISBN: 978-1-315-62071-8 (eBook PDF)

Table of contents

Hydraulic Engineering IV – Xie (Ed.)
© 2016 Taylor & Francis Group, London, ISBN 978-1-138-02948-4

Preface

Hydraulic research is developing beyond traditional civil engineering to satisfy increasing demands in natural hazards, structural safety assessment and also environmental research. In such conditions, this book embraces a variety of research studies presented at the 4th International Technical Conference on Hydraulic Engineering (CHE 2016), including the 5th International Workshop on Environment and Safety Engineering (WESE 2016) and the 2nd International Structural and Civil Engineering Workshop (SCEW 2016), being held in Hong Kong, July 16–17, 2016. The series of conferences was conceived and organized with the aim to promote technological progress and activities, technical transfer and cooperation, and opportunities for engineers and researchers to maintain and improve scientific and technical competence in the field of hydraulic engineering, environment and safety engineering, and other related fields.

38 technical papers were selected for publication in the proceedings. Each of the papers has been peer reviewed by recognized specialists and revised prior to acceptance for publication. The papers embody a mix of theory and practice, planning and reflection participation, and observation to provide the rich diversity of perspectives represented at the conference. This book review recent advances in scientific theories and modelling technologies that are important for understanding the challenging issues in hydraulic engineering and environment engineering. Some excellent papers were submitted to this book, and some highlights include:

- flood forecast and control, e.g. probability-based flood forecast, rogue wave.
- hydraulic structures, e.g. arch dam, new conceptual full tubular submersible gate-pump.
- stability and risk analysis of hydraulic structures, e.g. risk prioritization of hydropower dams, neural network for dam deformation, abutment instability of arch dam.
- hydraulic engineering technology, e.g. dredging techniques and rake tooth optimization, blending ground sand dust with unqualified fly ash in RCC, steel reinforced concrete in radial gate pier.
- geotechnical aspects, e.g. particle breakage and creep deformation prediction of rockfill materials, suffusion in a coarser soil matrix.
- fluid-structure interactions, e.g. nonlinear fluid-solid interaction of bridge rectangular pier in deep water.
- earthquake hazards, e.g. resistivity image as a new indicator for earthquakes.
- civil engineering, e.g. tunnel inflow change considering groundwater level drawdown, anchor bolt structure, cylindrical corrugated steel silos without columns, optimization of road intersection.
- concrete and construction materials, e.g. concrete anti-clay additives, new Modified Polymer Concrete (MPC) with high strength and light weight using fly-ash.
- pollution and control, e.g. effect of an Enhanced Ecological Floating Bed (EEFB) on phytoplankton community in an urban tidal river, benthic diatom communities in Pearl River (China), hydrothermal synthesis of gypsum whisker by industrial phosphogypsum washed with water.
- water resources and water treatment, e.g. contamination of dredged sediment in Taihu Lake (China), transport of sediment and heavy metal in a bay system, anaerobic co-digestion of waste activated sludge and waste wine distillate.
- environmental fluid dynamics, e.g. simplified simulation of flows with turbulent macrostructure.

- safety analysis in engineering, e.g. effects of fire locations on the evacuation of high-rise building, Risk-Based Inspection (RBI) in atmospheric storage tanks.
- dust control, e.g. dust removal efficiency of gas-water nozzle, dust control technology in fully mechanized excavation face.
- modelling technology in hydraulic engineering, e.g. new explicit analytical solutions of rogue wave based on the HAM.
- numerical software and applications, e.g. neural network for dam deformation, simulation of flows with turbulent macrostructure, numerical simulation of knotless fishing nets under the action of water.
- other aspects, e.g. calculation of CO_2 emission from thermal power industry based on environmental statistics 2014 (China), public assembly occupancy based on risk assessment.

At the very time we would like to express our deep gratitude to all authors, reviewers for their excellent work, and Léon Bijnsdorp, Lukas Goosen and other editors from Taylor & Francis Group for their wonderful work.

Hydraulic Engineering IV – Xie (Ed.)
© 2016 Taylor & Francis Group, London, ISBN 978-1-138-02948-4

Particle breakage and creep deformation prediction of rockfill materials

Cheng-fa Deng
Zhejiang Guangchuan Engineering Consultation Co., Ltd., Hangzhou, China
Zhejiang Institute of Hydraulics and Estuary, Hangzhou, China

Xue-mei Li
Zhejiang Institute of Hydraulics and Estuary, Hangzhou, China
Zhejiang Provincial Key Laboratory of Hydraulic Disaster Prevention and Mitigation, Hangzhou, China

Lei-lei Zheng
Zhejiang Institute of Hydraulics and Estuary, Hangzhou, China

Yong-ming Wang
POWERCHINA Huadong Engineering Corporation Ltd., Hangzhou, China

ABSTRACT: Rockfill creep is mainly caused by particle breakage. This paper conducted the particle breakage experiment on rockfill materials using a large direct shear apparatus, introduced discarding parameters to analyze test results, and carried out 3D creep computation on a typical project. It is concluded that in the range of normal stress in study, there is a linear relationship between the discarding ratio/weighted discarding ratio and normal force with a significant turning point of slope. As the particle size decreases, the relationship between the discarding ratio and normal stress shows a linear tendency. When analyzing the rockfill creep, it was found that dam body settlement, horizontal displacement, slab deflection, and peripheral joint deformation increased by 46.3–150%, suggesting that rockfill creep has a great influence on the safety of Concrete Face Rockfill Dam (CFRD). However, as the dam is not high, there is little possibility that the studied dam would experience slab tension crack and standard-exceeding water stop, regardless of its large deformation during water storage period.

1 INTRODUCTION

Creep is one of the basic deformation characteristics of rockfill (LIANG Jun, 2003). The results for the prototype observation of several Concrete Face Rockfill Dams (CFRDs) suggest that during a certain period, the deformation of rockfill is increasing in a continuous and slow trend; for high-faced rockfill dams and those filled by mixing soft rock and soft–hard rock, the fluid deformation is even more obvious (ZHOU Wei, 2007) (SHAO Lei, 2013).

The rheological behaviors of rockfill are extremely complicated, closely related to mechanical phenomena and physical mechanisms. The lithology and quality, gradation feature and relative dense degree of rockfill materials, and limitation of external forces are preliminary conditions of rockfill rheology. Studies show that rockfill rheology is mainly manifested as breakage, refining, displacement, and realignment of particles in microscopic-perspective. Similar to secondary consolidation of soil, the deformation of rockfill materials are related to time, but in a finite way (HU Ying, 2007) (MA Gang, 2012) (MARSAL R J., 1967).

Conducting experimental study of particle breakage on rockfill materials is helpful to understand the deformation mechanism and lay a theoretical foundation for the rockfill creep model. In this paper, large-scale direct shear testing was carried out on tuff rockfill. Based on results of the test, splitting variables were analyzed to study the developing rule of particle breakage. This paper also performed 3D finite element calculation on the creep of CFRD, to predict the deformation, stress, and joint displacement of dams considering rockfill creep, which may provide reference to similar projects.

2 PARTICLE BREAKAGE TEST OF ROCKFILL MATERIALS

In general, the particle breakage before and after testing is quantitatively described by the gradation variation. Commonly used indexes are breakage ratio B_g proposed by Marsal (1967) and a relative breakage ratio proposed by Hardin (1985). Based on the existing tri-axial testing data and the internal relation of particle breakage and gradation variation, XU Ri-qing *et al.* (2013) put forward new concepts of discarding ratio R_k and weighted discarding ratio B_d for the crushing of a single fraction and granular soils as a whole, respectively. The two concepts could fully reflect the particle breakage of rockfill materials. The discarding ratio represents the percentage of discarded particle quality in the overall quality of original fraction after the crushing of fraction according to the granular diameter. The weighted discarding ratio represents the weighted average of discarding ratio Wk with granular content of fractions under original gradation as weight. The expression is as follows:

$$B_d = \frac{1}{100\%} \sum_{k=1}^{n} W_k R_k \tag{1}$$

where R_k ($k = 1, 2, 3, ..., n$) represents the discarding ratio of the kth fraction after crushing and B_d represents the weighted discarding ratio of coarse-grained soil.

Discarding parameters were introduced into the data processing of large-scale direct shear test, whose results are shown in Figure 1.

Figure 2 shows a linear relationship between the discarding ratio and normal stress; in addition, there is a significant turning point of slope in such a linear relationship, with the corresponding normal stress being at 300 kpa. When the normal stress is less than 300 kpa, the line is steeper, indicating that as the normal stress is increased, the discarding ratio increases rapidly and the particle breakage is relatively more obvious. When the normal stress is larger than 300 kpa, the line becomes flatter, indicating that as the normal stress is increased, the discarding ratio increases slowly and the particle breakage is relatively less obvious. With the decrease in particle size, the relationship of discarding ratio and normal stress shows a

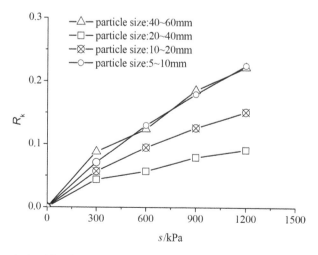

Figure 1. R_k—σ relationship of the coarse-grained soil with different fractions.

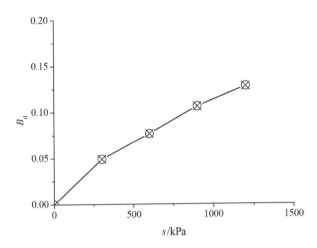

Figure 2. B_d—σ relationship of the coarse-grained soil.

linear tendency. Meanwhile, the particle breakage of fraction with a size range of 40–60 mm and 5–10 mm is more evident, but the fraction with a size range of 20–40 mm is less evident. The results are inconsistent with the general viewpoint that under the same conditions, the larger the particle, the more the possibility of breakage. This phenomenon can be explained by three reasons: 1) For the same coarse-grained soil with continuous gradation, its particle breakage is related to not only the normal force, the strength of mother rock, but also the particle shape, overall gradation composition and other factors. 2) The calculation of discarding ratio is based on the hypothesis of geometric progression distribution. With continuous effect of load, there will be certain discrepancy between the composition of fraction particles and the hypothesis. 3) The clastic materials after particle breakage may be categorized into smaller group, or partially be categorized into smaller group, or may even belong to the original group. Therefore, after the crushing, the granular contents of fractions may be greater, smaller, or unchanged compared to that of original gradation.

3 CONSTITUTIVE MODEL OF ROCKFILL CREEP

Under the stress such as deadweight (during construction) and water load (during operation), coarse grains of rockfill would experience crushing, refining, rearrangement of particles, and displacement and filling pores. These phenomena manifest as slow deformation in a macroscopic sense, i.e. creep. Initially, the creep rate is fast, depending on the rate of filling pores, and when the rate of filling and rearranging slows down, the creep rate of rockfill becomes slow. Therefore, establishing a creep model for rockfill materials is to reasonably simulate the deformation process of rockfill over time (WANG Yong, 2000). SHEN Zhu-jiang (1998), Academician of Chinese Academy of Sciences, proposed a three-parameter model of rockfill creep:

$$\varepsilon(t) = \varepsilon_f \left(1 - e^{-ct}\right) \tag{2}$$

where ε_f is the final creep volume when $t \to \infty$ and c is the proportion of creep volume to ε_f when $t = 0$.

As the rules describing volume creep and shearing creep are different, two equations are suggested:

$$\varepsilon_{vf} = b(\frac{\sigma_3}{Pa})^n \tag{3}$$

3

$$\varepsilon_{sf} = d\frac{SL}{1-SL}$$

(4)

where b is the final volume creep with σ_3 being the atmospheric pressure, n is the power of volume creep changing with confining pressure σ_3, d is the final shearing creep when stress level $SL = 0.5$; and at the time of breakage, $SL = 1.0$, $\varepsilon_{sf} \to \infty$. If $SL \geq 1.0$ at computing, then $SL = 0.95$.

4 3D CREEP ANALYSIS

The CFRD of Shuangxikou Reservoir is located at the Dayin Creek, a tributary of Yaojiang River, Yuyao, Zhejiang Province. The reservoir is built on a wide valley with a total storage capacity of 33.98 million m³, the largest dam height of 67.10 m, the width at dam top of 6.9 m, and a length of 426.0 m.

To speed up the construction, the filling of dam was completed in a short period of time followed by concrete face pouring without reserving time for settlement, which would lead to serious deformation after water storage and jeopardize dam safety. To predict deformation and stress after water storage, the paper conducted computational analysis on rockfill creep. The 3D creep model of the dam is shown in Figure 3.

The constitutive model of dam building materials adopts Duncan E-B model, and parameters are obtained from inversion (DENG Cheng-fa, 2013), as shown in Table 1. The parameters are $b = 0.0004$, $c = 0.007$, and $d = 0.006$.

Table 2 shows the displacement, stress, and deformation differences of dams during the normal water storage period with and without consideration of creep. Figures 4 and 5 show contour line of rockfill settlement and horizontal displacement, respectively, considering creep. Figures 6 and 7 show contour line of stress.

From Tables 1 and 2 and Figures 4–7, it can be suggested that creep exerted a large influence on deformation and stress of dams; in particular, the differences in dam body settlement, horizontal displacement, slab deflection, peripheral joint deformation reached 39.1–60%, and the difference in stress along slope was 31.7%. The ranges increased from 46.3 to 150% compared to that without considering creep.

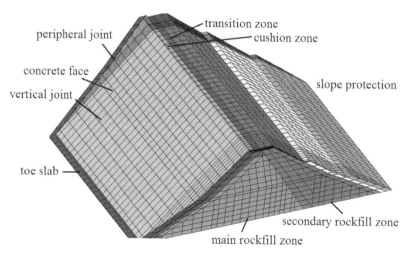

Figure 3. 3D overall grid.

Table 1. E-B model parameters of dam building materials.

Type of material	$\gamma_d/g \cdot cm^{-3}$	$\varphi_0/°$	$\Delta\varphi/°$	R_f	K	n	K_b	m
Cushion material	2.25	52.0	9.0	0.75	1,100	0.35	420	0.21
Transition material	2.22	51.0	8.0	0.75	980	0.28	380	0.20
Main rockfill material	2.20	55.0	10.0	0.83	970	0.30	350	0.19
Secondary rockfill material	2.05	52.0	13.0	0.78	790	0.40	330	0.22

Table 2. Computation results of creep during the water storage period.

Difference item	Considering creep	Without considering creep	Difference proportion/%	Increasing range/%
Max settlement (cm)	32.02	19.5	39.1	64.2
Max horizontal deformation (cm)	8.75	4.73	45.9	85.0
Slab deflection (cm)	14.79	6.56	55.6	125.5
Compressive stress along slope (MPa)	3.57	2.44	31.7	46.3
Tensile stress along dam axis (MPa)	−0.86	−0.71	17.4	21.1
Compressive stress along dam axis (MPa)	1.36	1.22	10.3	11.5
Peripheral joint settlement (mm)	5	2	60.0	150.0
Peripheral joint shearing (mm)	12	5	58.3	140.0
Peripheral joint confinement (mm)	10	4	60.0	150.0

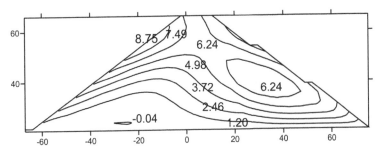

Figure 4. Contour graph of horizontal displacement considering creep (cm).

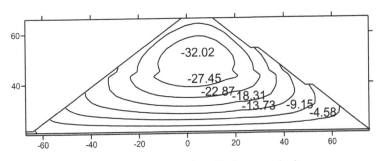

Figure 5. Contour graph of vertical displacement considering creep (cm).

5 CONCLUSION

Creep is one of the basic deformation characteristics of rockfill, which is mainly caused by particle breakage of coarse-grain materials. This paper utilized a large direct shear apparatus to carry out the particle breakage experiment on rockfill materials, introduced the discarding parameters to quantitatively analyze the breakage level, and carried out 3D creep computation on a typical project. It can be concluded that:

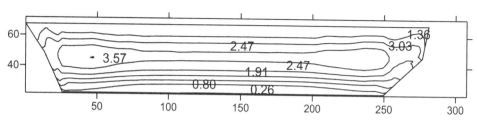

Figure 6. Contour graph of stress along slope considering creep (MPa).

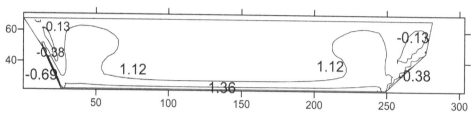

Figure 7. Contour graph of stress along axis considering creep (MPa).

1. In the range of normal stress under study, there is a linear relationship between the discarding ratio/weighted discarding ratio and normal force, and there is a significant turning point of slope in such relationship. With the decrease in particle size, the relationship between the discarding ratio and the normal stress shows a linear tendency.
2. Compared to the conditions without consideration of creep, when taking creep into account, dam body settlement, horizontal displacement, slab deflection, and peripheral joint deformation increased by 46.3–150%, suggesting that rockfill creep has a great influence on CFRD safety. However, with a medium dam height, the studied dam is still safe with a relatively small possibility of having slab tension crack and standard-exceeding water stop, regardless of large deformation during the water storage period.

REFERENCES

Deng Cheng-fa. Impact study of concrete face rockfill dam settlement on engineering safety and acceptance[R]. Zhejiang Institute of Hydraulics & Estuary, 2013.

Hardin C S. Crushing of soil particles[J]. Journal of Geotechnical Engineering,1985,111(10):1 177–1 192.

Hu Ying, YAN Shengcun. Back Analysis and Application of Creep Parameters for Rockfill[J].Water Power, 2007,33(8):19–21.

Liang Jun, Liu Han-long, Gao Yu-fong. Creep mechanism and breakage behaviour of rockfill[J]. Rock and Soil Mechanics, 2003, 24(3): 479–483.

MA Gang, Zhou Wei, Chang Xiao-lin, et al. Mesomechanically numerical simulation of rockfill rheology based on particle deterioration[J]. Rock and Soil Mechanics, 2012,33(Suppl):257–264.

Marsal R J. Large-scale testing of rockfill materials[J]. Journal of the Soil Mechanics and Foundations Division,1967,93(2):27–43.

Shao Lei, Chi Shi-chun, Wang Zhen-xing. Rheological model for rockfill based on sub-critical crack expansion theory[J]. Chinese Journal of Geotechnical Engineering, 2013,35(1):66–74.

Shen Zhu-jiang, Zhao Kui-zhi. Back analysis of CFRD creep[J]. Journal of Hydraulic Engineering, 1998, 6(6):1–6.

Wang Yong. Analysis on rheology mechanism and study method of rockfill[J]. Chinese Journal of Rock Mechanics and Engineering, 2000, 19(4):526–530.

Xu Ri-qing, Chang Shuai, Li Xuegang, et al. A quantization method for crushing of granular soil based on discarding parameters[J]. Chinese Journal of Geotechnical Engineering,2013,35(12): 2 179–2 185.

Zhou Wei, Chan g Xiaolin, HU Ying, et al. Creep analysis of high concrete face rockfill dam[J]. Engineering Journal of Wuhan University, 2005,38(6):96–99.

Zhou Wei, Hu Ying, Yang Qigui, et al. Study on creep mechanism and long-term deformation prediction for high concrete face rockfill dam[J]. SHUILI XUEBAO, 2007,(Supp):100–104.

Hydraulic Engineering IV – Xie (Ed.)
© 2016 Taylor & Francis Group, London, ISBN 978-1-138-02948-4

An energy efficiency evaluation measure of water supply systems for high-rise buildings

L.T. Wong, K.W. Mui & Y. Zhou
*Department of Building Services Engineering, The Hong Kong Polytechnic University,
Hong Kong, China*

ABSTRACT: High-rise housing, a trend in densely populated cities around the world, increases energy use for water supply and corresponding greenhouse gas emissions. This paper presents an energy efficiency evaluation measure for water supply system designs and demonstrates its applications in a typical water supply system installation of buildings. The energy efficiency for water supply system operation is the potential energy of the water demands in buildings divided by the required pumping energy. To demonstrate that the measure is useful for establishing optimal design solutions that integrate energy consumption into urban water planning processes that cater to various building demands and usage patterns, measurement data of high-rise residential buildings in Hong Kong were used. The results showed that this measure could serve as a useful benchmark reference for water supply system designs and establish demand management programs for buildings.

1 INTRODUCTION

Very tall buildings are a trend in recent developments in Hong Kong, where is a developed city on hilly terrain with limited usable land for buildings. It has been estimated that the current average height of the residential building in the city is estimated to be 26 storeys (Cheng, Yen, Wong & Ho, 2008). For high-rise buildings, gravity storage tanks on building rooftops (or on intermediate mechanical floors) are designed in order to distribute water through down-feed pipes (Wong & Mui, 2007) and pressure reducing valves with adjustable settings, and screwed joints are commonly installed to minimize the problems of water leakage or damage in supply pipes and appliances caused by excessive water pressure on lower floors in low demand situations.

An energy efficiency evaluation measure is proposed for water supply system designs in buildings, with verification measurements in some high-rise residential buildings of Hong Kong (Cheung, Mui & Wong, 2012; Cheung Mui & Wong, 2013). The results showed that the energy efficiency of many existing high-rise water supply systems was about 0.25 and could be improved significantly via relocation of water storage tank. The corresponding annual electricity that could have been saved was 160–410 TJ, a 0.1–0.3% of the total annual electricity consumption in the city. The parameter was further tested to be applicable for various building demands of non-uniform usage patterns. The results showed that the energy efficiency of some high-rise water supply systems was up to 0.24 and could be as high as 0.34 in order to avoid over pressure at the water demand points (Wong, Cheung & Mui, 2013).

2 ENERGY EFFICIENCY

The energy efficiency of the water supply system in high-rise buildings is defined as the required potential energy E_{out} at the demand location divided by the pumping energy E_{pump} of the supply system. It can be determined using the system heights, pipe friction, and allowable pressure head, as shown in Figure 1, for example, roof tank water supply system for maintaining water supply appliance operation (Cheung, Mui & Wong, 2012):

$$\alpha = \frac{E_{out}}{E_{pump}} \qquad (1)$$

E_{out} (MJ) is the potential energy for the volumetric water demands v_i at height h_i, which is given as follows:

$$E_{out} = \rho g \sum_{i=1}^{n} v_i h_i \qquad (2)$$

where ρ is the water density (= 1000 kgm^{-3}) and g is the gravity (= 9.81 ms^{-2}).

The pumping energy of lifting water up from the break tank to the roof tank E_{pump} (MJ) is defined below, where η_c is the design overall transmission efficiency; h_l is the height difference between the break tank water surface and the roof tank inlet, which is the sum of the height measured from the roof tank base to the tank inlet h_c, the height difference between the demand n and the tank base h_b, and the height difference between the break tank water surface and the top demand location h_n, H_o is the desired minimum water pressure head assumed at the roof tank inlet point and H_f is the friction head required in the up-feed water pipe:

$$E_{pump} = \frac{\rho g \left(h_l \sum_{i=1}^{n} v_i + H_f + H_0 \right)}{\eta_c}; \quad h_l = h_c + h_b + h_n \qquad (3)$$

The energy per volumetric water consumption e is an indicator of engineering design and is expressed as follows:

$$e = \frac{E}{v}; \quad v = \sum_{i=1}^{n} v_i \qquad (4)$$

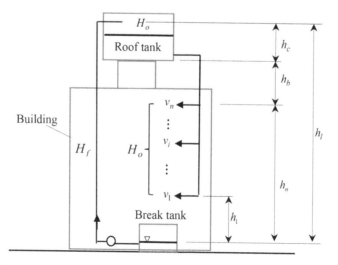

Figure 1.   Roof tank water supply systems in buildings.

It is noted for the design overall transmission efficiency η_c (34–65%) accounted for 50–80% of the pump efficiency η_p, about 90% of the mechanical transmission efficiency η_m and 70–90% of the electric motor efficiency η_e:

$$\eta_c = \eta_p \eta_m \eta_e \tag{5}$$

3 APPLICATION EXAMPLES

An example of water supply system for high-rise tank with design parameters is presented in Table 1. It is water supply system for high-rise tank for 600 residential WC cisterns (Mui and Wong, 2013). Design inflow rates were determined for the design and installed conditions and then were compared with a case using some existing design practices (Wong, Mui, Lau & Zhou, 2014; Plumbing Services Design Guide, 2002). The installed condition allowed a safety margin of 30%, and a larger pump was selected. Its roof tank was fed by a pump at the design flow rates through a 67-mm diameter pipe. The total static head for $h_l = 100$ m was counted and frictional head loss H for equivalent pipe length of $h_{fo} = 150$ m was included. To determine the system efficiencies, an average height of demand locations $h_d = 40$ m and an overall pump efficiency $\eta_c = 0.5625$ (Wong, Mui, Lau & Zhou, 2014) were applied.

$$\alpha = \frac{H_d \eta_c}{H_{fo} + h_l} \tag{6}$$

To determine the pumping energies, the following equation was used:

$$E_{pump} = \frac{\rho g V \left(H_{fo} + h_l \right)}{\eta_c} \tag{7}$$

Table 1 shows that system efficiency of the installed system is reduced from 0.2123 to 0.1883, corresponding to an efficiency drop of 12 or 13–14% more energy consumption. It is due to the additional frictional loss at the supply line of the higher water velocity as compared with the design one.

Proper commissioning for the pump will be solution to achieve the design flow rate; therefore, a variable speed control is required. Based on a mass balance analysis on the storage tank, the relationship between the inflow rate to the tank and the storage tank size is

Table 1. An example of high-rise tank water supply system for 600 residential WC cisterns (static head = 100 m; a 67-mm diameter feed pipe with equivalent length = 150 m).

Parameters	Design	Installed	Variable speed pump	Intermediate tank
Tank size (m³)	0.25–27	0.25–27	0.25–27	(2 nos.) 0.25–13.5
Daily consumption (m³)	76–82	76–82	76–82	76–82
Design inflow rate (Ls⁻¹)	5.1	11.2	0.95–1.9	3.2
Feed pipe water velocity (ms⁻¹)	1.5	3.1	<0.5–0.7	0.9
Friction head loss (m)	6	19.5	0.33–1.2	1.4–2.7
System efficiency	0.2123	0.1883	0.2223–0.2243	0.2735
Electricity power (kW)	9.5	23.34	1.66–3.35	8.7
Daily pumping energy (kWh)	39–42	44–48	37–40	29–32

determined for the usage patterns (Mui and Wong, 2013). A simulation showed that the daily consumption is from 76 to 82 m³, and the design flow rate is from 0.95 L/s (at a tank size of 27 m³) to 1.9 L/s (at a tank size of 0.25 m³) to fulfill the water demands in some residential buildings (Wong & Mui, 2008). The system efficiency of the supply system of a variable speed pump can be up to 0.2243, corresponding to saving 5% of pumping energy as compared with the design one. In other words, energy loss due to the system friction in high-rise pumping can be significant, and the system friction optimization should not be ignored in future designs of the tank water supply network.

Zoning a high-rise water supply system with an intermediate tank would reduce pumping energy due to a better pressure control (Cheung, Mui and Wong, 2013). The saving can be indicated by the higher system efficiency value of 0.2735, corresponding to a 24–26% energy saving, as indicated in Table 1.

4 CONCLUSION

With particular emphasis on improving energy efficiency in water supply systems, this paper proposed the energy efficiency of the water supply system calculated by the potential energy of the water at the demand location divided by the required energy of the system. The values of system efficiency were determined, for example water system and some engineering solutions addressing the water demands. The results showed that this measure could serve as a useful benchmark reference for the designs of water supply system and establish demand management programs for buildings.

ACKNOWLEDGMENT

This work was partially supported by a grant from the Research Grants Council of the HKSAR, China (PolyU5272/13E).

REFERENCES

Cheng C.L., Yen C.J., Wong L.T. & Ho K.C. 2008. An evaluation tool of infection risk analysis for drainage system in high-rise residential buildings. *Building Services Engineering Research and Technology* 29(3):233–248.

Cheung C.T., Mui, K.W. & Wong L.T. 2012. Energy implications for water supply tanks in high-rise buildings. The 38th CIBW062 International Symposium of Water Supply and Drainage for Buildings, 27–30 August, Edinburgh, Scotland, UK.

Cheung C.T., Mui K.W. & Wong L.T. 2013. Energy efficiency of elevated water supply tanks for high-rise buildings. *Applied Energy* 103(3):685–691.

Mui K.W. & Wong L.T. 2013. Modelling occurrence and duration of building drainage discharge loads from random and intermittent appliance flushes. *Building Services Engineering Research and Technology* 34(4):381–392.

Plumbing Services Design Guide 2002. The Institute of Plumbing. England: Hornchurch.

Wong L.T. & Mui K.W. 2007. Modeling water consumption and flow rates for flushing water systems in high-rise residential buildings in Hong Kong. *Building and Environment* 42(5):2024–2034.

Wong L.T. & Mui K.W. 2008. Stochastic modelling of water demand by domestic washrooms in residential tower blocks. *Water Environment Journal* 22:125–130.

Wong L.T., Cheung C.T. & Mui K.W. 2013. Energy efficiency benchmarks of example roof-tank water supply system for high-rise low-cost housings in Hong Kong. The 39th CIBW062 International Symposium of Water Supply and Drainage for Buildings, 17–20 September, Nagano City, Japan.

Wong L.T., Mui K.W., Lau C.P. & Zhou Y. 2014. Pump efficiency of water supply systems in buildings of Hong Kong. *Energy Procedia* 61:335–338.

Hydraulic Engineering IV – Xie (Ed.)
© 2016 Taylor & Francis Group, London, ISBN 978-1-138-02948-4

Effect of the load history of hydraulic gradient on suffusion

Y.L. Luo & X.F. Kong
College of Water Conservancy and Hydropower Engineering, Hohai University, Nanjing, China

ABSTRACT: This paper presented a series of hydro-mechanical suffusion experiments to study the effect of the load history of hydraulic gradient on the evolution of suffusion. Four different increase velocities of hydraulic gradient were used to simulate the different load histories of hydraulic gradient. The results indicate that the load history of hydraulic gradient has a significant influence on the evolution of suffusion. The faster the increase velocity of hydraulic gradient, the smaller the suffusion failure hydraulic gradient, and the larger the cumulative eroded mass. A fast increase velocity of hydraulic gradient is harmful for the safety of hydraulic structures. The results presented here explain the reason why suffusion always happens when the water level at the upstream side rises quickly during flood season.

1 INTRODUCTION

Suffusion, a type of internal erosion, refers to the migration of fine grains in a coarser soil matrix. With the loss of fine particles, local cavities and differential settlements may occur. Suffusion may be affected by hydraulic and stress conditions. Sterpi (2003) studied the influence of hydraulic gradient on suffusion and proposed an empirical law expressing the quantity of eroded particles with hydraulic gradient and time. Bendahmane et al. (2008) designed a triaxial suffusion apparatus and carried out a series of experiments to investigate the effect of hydraulic gradient, clay content, and confining pressure on suffusion. The spatial and temporal progressions of suffusion were studied by Moffat et al. (2011). Li and Fannin (2012) pointed out that suffusion was related to hydraulic gradient and effective stress in the finer fraction. Richards and Reddy (2012) developed a true triaxial apparatus and found that the hydraulic velocity was a more reliable indicator of suffusion than hydraulic gradient for non-cohesive soils. The effect of stress state on critical hydraulic gradient for suffusion was investigated by Chang and Zhang (2013), and three critical gradients termed as initiation, skeleton-deformation, and failure hydraulic gradients were defined. It can be seen that the present studies mainly focus on the influences of hydraulic gradient and stress on suffusion, while the influence of load history of hydraulic gradient on suffusion was not studied, which is significantly critical for investigating the suffusion mechanism during flood season. This paper used the increase velocity of hydraulic gradient to simulate the different load histories of hydraulic gradient, and then four hydro-mechanical coupling suffusion experiments under different increase velocities of hydraulic gradient were performed. Finally, the effect of load history of hydraulic gradient was evaluated.

2 METHODOLOGY

2.1 *Apparatus*

The hydro-mechanical coupling suffusion apparatus, depicted in Figure 1, can be used to study the evolution of suffusion under different triaxial stress states and hydraulic gradients. It mainly consists of a confining pressure system, an axial loading system, a seepage pressure system, a funnel-shaped drainage system, and a data acquisition system. The confining

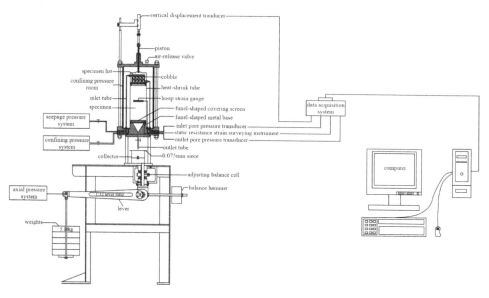

Figure 1. Schematic diagram of the new apparatus.

and axial pressure systems were used to simulate the stress state of specimen. The largest confining pressure and axial load are up to 2.0 MPa and 30 kN, respectively. The seepage pressure system provided seepage for fine particle migration. The total head difference across specimen was controlled by a water tank with adjustable height. The funnel-shaped draining system was designed to allow fine particle to freely exit specimen. The data acquisition system could monitor settlement, volume change and total head difference of the specimen. The flow rate was measured manually. The detailed information on this apparatus can be found in Luo et al. (2013).

2.2 Characteristics of experimental soil

The soil tested here is an internally unstable sandy gravel according to the geometric criteria of internal instability (Kenney and Lau 1985). The Particle Size Distribution (PSD) is shown in Figure 2. The soil can be classified as SP based on unified soil classification system. The height and diameter of the specimen were both 10 cm, and the initial porosity and dry density were 0.301 and 1.91 g/cm³. It should be pointed out that the boundary effects can be eliminated in the hydro-mechanical experiments, because the lateral surface of the specimen is compressed firmly by confining pressure.

2.3 Experimental procedures

1. Specimen preparation and saturation. First, moderate masses of dry soil and water were prepared and mixed completely, and then the mixture was divided into five parts. Second, the five parts were compacted separately. Third, the compacted specimen was fixed on the base of confining pressure cell. The outlet tube, inlet tube and inlet piezometer tube were turned off, and then the specimen was saturated using the outlet piezometer tube under a low pressure. Finally, the outlet piezometer tube was turned off when the specimen was saturated completely, and then the moveable water tank was connected to the inlet tube.
2. Application of confining pressure. The confining pressure cell was covered, and the air-release valve was opened. Water was added into the cell until it was full. The inlet

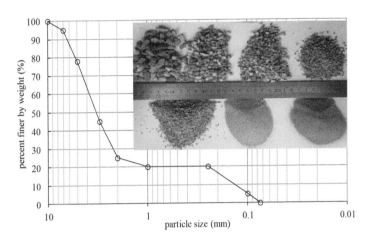

Figure 2.   Particle size distribution of sandy gravel.

piezometer tube was opened, and then the confining pressure output control gauge was turned on to apply confining pressure gradually. The confining pressure was 0.4 MPa in this paper.

3. Application of seepage pressure. The outlet tube, inlet tube and outlet piezometer tube were turned on, and the seepage pressure was applied by raising the water tank, then a downward seepage flowed into the specimen, and discharged from the funnel-shaped drainage system. The inlet and outlet hydraulic heads, flow rate, eroded mass and time were recorded The hydraulic gradient was calculated by the ratio of the difference between the inlet and outlet hydraulic heads to the height of specimen.

The critical suffusion hydraulic gradient was the minimum hydraulic gradient at which fine particles start to migrate out of specimen. Once the critical suffusion hydraulic gradient reached in every experiment, then four different increase velocities of hydraulic gradient, 0.2, 0.4, 0.6, and 0.8, were applied in the subsequent experiments, respectively.

3   RESULTS AND DISCUSSION

3.1   *Effect on suffusion failure hydraulic gradient*

The critical suffusion hydraulic gradients for the four experiments under different increase velocities of hydraulic gradient were 0.385, 0.415, 0.42, and 0.4, respectively, they were significantly consistent, which indicate that the four experiments had the same initial experimental conditions and had good repeatability.

Suffusion failure was defined when the cumulative eroded mass increased suddenly For the four experiments with the increase velocities of hydraulic gradient, 0.2, 0.4, 0.6, and 0.8, the corresponding failure hydraulic gradients were 2.835, 2.1, 2.21, and 1.965, respectively, as shown in Figure 3. It can be seen that the increase velocity of hydraulic gradient has a significant influence on the suffusion failure hydraulic gradient, and the faster the increase velocity of hydraulic gradient, the smaller the suffusion failure hydraulic gradient. A fast increase velocity of hydraulic gradient is harmful for the safety of hydraulic structure suffering from suffusion.

3.2   *Effect on cumulative eroded mass of fine particles*

Figure 3 also depicts the relationships between hydraulic gradient and cumulative eroded mass. It can be seen that the increase velocity of hydraulic gradient significantly affect the

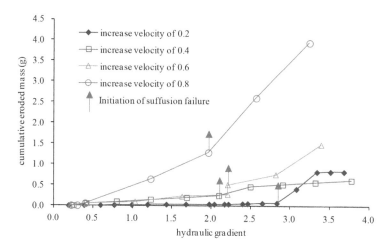

Figure 3.   Variation of suffusion failure hydraulic gradient.

cumulative eroded mass of fine particles. The cumulative eroded mass at the initiation of suffusion failure and at the end of the experiment under the increase velocity of hydraulic gradient 0.2 was 0.075 g and 0.835 g, respectively. Similarly, that under the increase velocity of hydraulic gradient 0.8 was 1.284 g and 3.949 g, respectively, which were approximately 17 and 5 times of those under the increase velocity of hydraulic gradient 0.2.

3.3   Effect on permeability of soil

Figure 4 shows the evolution of permeability of soil under different increase velocities of hydraulic gradient. It can be seen that the variation of permeability can be split into three stages. In Stage 1, the faster the increase velocity of hydraulic gradient, the smaller the permeability. The reason may be that a great number of fine particles, which started to migrate under large increase velocity of hydraulic gradient, clogged the pores of specimen. Stage 2 is a transit stage, where the permeability of the specimen under slow increase velocity of hydraulic gradient started to decrease owing to the clogging of pores, while that under fast increase velocity of hydraulic gradient started to increase owing to the reopening of the clogging pores. During Stage 3, massive fine particles were blowing out of the specimen under fast increase velocity of hydraulic gradient owing to the entire blowing out of the clogging

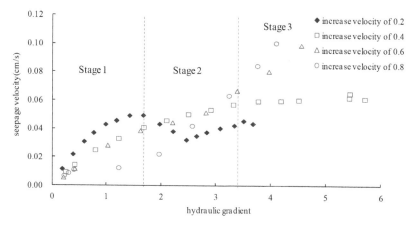

Figure 4.   Evolution of permeability under different increase velocities of hydraulic gradient.

pores. For the specimen under the increase velocity of hydraulic gradient of 0.2, the permeability in Stage 1 and 3 was 0.0541 and 0.0129, respectively. Similarly, for the specimen under the increase velocity of hydraulic gradient of 0.8, the permeability in Stage 1 and 3 was 0.0286 and 0.0431, respectively.

The evolution of permeability under different increase velocities of hydraulic gradient also indicates that the increase velocity of hydraulic gradient has great influence on suffusion. A fast increase velocity, which is harmful for the safety of hydraulic structures, will significantly increase the permeability at the failure of suffusion, while a slow increase velocity will significantly decrease the permeability at the failure of suffusion.

4 CONCLUSIONS

This paper performed four hydro-mechanical suffusion experiments under different increase velocities of hydraulic gradient to study the effect of the load history of hydraulic gradient on suffusion. The experimental results indicate that the load history of hydraulic gradient significantly influence the evolution of suffusion. A fast increase velocity of hydraulic gradient significantly decreases the suffusion failure hydraulic gradient, increases the cumulative eroded mass of fine particles, and increases the permeability at the failure of suffusion. The evolution of suffusion is a complicated and iterative process involving fine particles migration, pores clogged, blowing out of the clogged pores and fine particles remigration. A fast increase velocity of hydraulic gradient is harmful for the safety of hydraulic structures, which explains the reason why suffusion always happens when the water level at the upstream side rises quickly during flood season.

REFERENCES

Bendahmane, F., Marot, D. & Alexis, A. 2008. Experimental parametric study of suffusion and backward erosion. Journal of Geotechnical and Geoenvironmental Engineering 134(1), 57–67.

Chang, D.S., & Zhang, L.M. 2013. Critical hydraulic gradients of internal erosion under complex stress states. Journal of Geotechnical and Geoenvironmental Engineering 139(9), 1454–1467.

Kenney, T.C. & Lau, D. 1985. Internal stability of granular filters. Canadian Geotechnical Journal 22(2), 215–225.

Li, M. & Fannin, R.J. 2012. "A theoretical envelope for internal instability of cohesionless soil." Geotechnique 62(1), 77–80.

Luo, Y.L., Qiao, L., Liu, X.X., Zhan, M.L. & Sheng, J.C. 2013. "Hydro-mechanical experiments on suffusion under long-term large hydraulic heads." Natural Hazards 65(3), 1361–1377.

Moffat, R.M., Fannin, R.J. & Garner, S. J. 2011. "Spatial and temporal progression of internal erosion in cohesionless soil." Canadian Geotechnical Journal 48(3), 399–412.

Richards, K.S. & Reddy, K.R. 2012. "Experimental investigation of initiation of backward erosion piping in soils." Geotechnique 62(10), 933–942.

Skempton, A.W. & Brogan, J.M. 1994. "Experiments on piping in sandy gravels." Geotechnique 44(3), 449–460.

Sterpi, D. 2003. "Effect of the erosion and transport of fine particles due to seepage flow." International Journal of Geomechanics 3(1), 111–122.

Hydraulic Engineering IV – Xie (Ed.)

Structural analysis and optimal design of full-tubular submersible gate pump

Jun Ding & Guohua Shen
Jiangsu Surveying and Design Institute of Water Resources Co., Ltd., Yangzhou, China

Haitian Huang
Water Resources Department of Jiangsu Province, Nanjing, China

ABSTRACT: In order to reveal the structural characteristics of full-tubular submersible gate pump, correct its layout and improve the calculation theories of structural stress, this paper expounds the structure and layout of the submersible gate pump. Based on the *Solid-Works* design software platform and using 3D *SimulationXpress* finite element analysis software, the analysis and calculation of stress and strain of the inlet gate were carried out. The advantages of the full-tubular submersible pump are that it has a shorter body length, is lightweight, has a diversified arrangement as well as is easy to combine with the gate and lift. According to the 3D finite element calculation theory, the simulated condition is close to the actual working condition, while the calculation results of gate stress and deformation are superior to the graphic methods. In a low-head irrigation and drainage area, such as polders along the Yangtze River and low-lying lands along the Huaihe River and Huaibei Plain in China, a lot of drainage culverts and sluice gates will be rebuilt into full-tubular submersible gate pumps. As a result, the structural analysis and optimal design methodology can provide a reliable theoretical support and a guarantee for a safe and reliable operation of the project.

1 INTRODUCTION

In order to solve the problems of the gravity flow discharge of the water flooding and water-logging in polder areas along the Yangtze River, low-lying areas, and suburbs in China, many gravity flow drainage culverts and small water checks have been built. Nowadays, changes in water regimen and the need for the so-called "clear and fresh water" in cities cannot be denied. Under the condition of outer river jacking, the gravity flow drainage culverts and small sluices are unable to discharge the flooding water and exchange water; thus, pumping stations must be built. However, new pumping stations have some disadvantages including shorter operational time and lower frequency usage, resulting in a waste of investment and ability to a certain extent. Based on the application of foreign countries and the actual situations in China, the combination technology of the gate and the submersible pump is put forward, namely the gate pump technology (Wang *S.L.*, 2009).

Considering the case of domestic and foreign research and applications, the environmental management and water conservancy project construction have been established relatively earlier in developed countries such as the USA, Japan, and European countries, and the gate pump technology has already been a more systematic application in these countries. In China, however, application of the gate pump technology has been started relatively late. In this respect, the research is mainly focused on the comparison between the installation methods of the middle- and the small-sized axial flow pumps, the submersible pump, and the gate pump together with the development of the submersible pump as well as the comparison between the connection modes (Jing *L.*, 2015). So far, no research has been conducted on the structure and design optimization of the submersible pump. Based on the applicationcondition, hydrology, and installation and connection methods, it is important to carry out the stress analysis and optimization design of the structure, connection types, and the opening and closing of the gate pump for a safe and reliable operation in theory and application.

2 STRUCTURE AND CHARACTERISTICS OF THE SUBMERSIBLE GATE PUMP

2.1 Classification of the submersible gate -pump

According to the arrangement types of the main shafts, submersible gate pumps are classified into vertical axial and horizontal tubular gate pumps. While considering the installation types of submerged motor structures, horizontal tubular gate pumps can be classified into semi-tubular submersible and full-tubular submersible gate pumps. The specified classifications of the gate pumps are shown in Figure 1.

2.2 The structural characteristics of the commonly used submersible gate pump

1. Vertical submersible axial flow gate pump
   Pit shaft, siphonium suspension, and pit shaft type of the prefabricated concrete are the main installation methods for the vertical submersible axial flow pumps: the most commonly used type is the pit shaft suspension (Rong Z.F., 2004). It has a strong adaptability to the change in the water level, especially suitable for the occasion of big fluctuation of the water level (Zhang et al., 2004, Liu et al., 2015). However, in the lower design head, there is large hydraulic loss, low efficiency, big motor power, and high operation cost.
   The submersible axial flow gate pump is the combination of a submersible axial flow pump and a gate. The submersible axial flow pump has a vertical intake and a horizontal discharge, and the gravity center of the pump is located in the tubular pump body. In order to reduce the vibration of the gate pump and to ensure the stability of the structure, a supporting beam is set up in the position of the pump body, and the pump is hung over the supporting beam without being fixed. When the integral connection and lifting are adopted for the gate pump, the bending moment produced becomes larger, so its effects on the gate structure are larger at the same time. Thus, the pump and the gate can only be connected by the self-coupling approach, and the gate and the submersible pump lifting must be carried out separately. The specific layout is shown in Figure 1.
2. Semi-tubular submersible gate pump
   The bulb tubular pump is a general use of the submersible pump (Wu et al., 2012). According to the position of the bulb body, it can be divided into a front or rear bulb type. The commonly used type is the rear bulb type, also known as a semi-tubular submersible pump, which is widely used both domestic and abroad. Under the condition of low head, the efficiency of the type is significantly higher than the vertical axial flow pump (Li et al., 2007). Because the axial size of the semi-tubular pump is larger, the heat dissipation condition of the unit in the bulb is poor, and the bulb will increase the flow resistance.

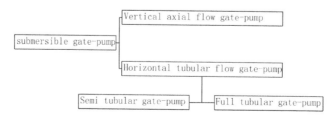

Figure 1. The classification of the submersible gate pump.

18

Semi-tubular submersible gate pump is a combination of a semi-tubular submersible pump and a gate, and the electric motor is arranged in the bulb body behind the impeller. Concerning the relative position between the pump and the gate, they can be classified into two types. One is the inlet of the pump is connected to the gate and the flap valve is away from the gate, similar to the layout shown in Figure 3 (a), while another is the outlet of the pump that is connected to the gate, and the flap valve and the gate are connected together and the pump inlet is away from the gate, similar to the layout shown in Figure 2 (b). Due to the longer pump shaft of the semi-submersible pump and the heavy weight of the pump body, the pump body can only be supported on the bottom plate. If the fixed connection is used, the influence on the gate structure is larger than that on the submersible axial flow gate pump. Therefore, only self-coupling connection with the gate can be adopted. In addition, the gate hoist and the submersible pump lifting must be carried out separately. The specific arrangement is shown in Figure 2.

2.3 Structural characteristics of all tubular submersible gate pumps

The full-tubular submersible gate pump has the advantages of the submersible pump and the full-tubular pump, and avoids the disadvantages and the complex degree of other pumping stations, which is the mechanical and electrical integration technology (Gu Z.S., 2010). The pump impeller is installed in the inner cavity of the motor, and the rotor is turned into the impeller of the water pump, where the invalid part of the motor is turned into a working part. The water flows across the inner cavity of the rotor of the submersible motor in a straight line. This type has the characteristics of compact structure, convenient installation, low noise, excellent heat diffusion, etc. while the characteristics of the tubular pump are maintained, such as superior hydraulic performance, smooth flow, and high system efficiency.

The combination of the full-tubular submersible pump and the gate is called the full-tubular submersible gate pump. The full-tubular submersible pump is the impeller built-in type, where the impeller and the motor rotor make up a whole, so that the axial length is shorter, the weight is lighter, and it is easy to combine with the gate. Based on the different

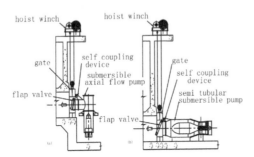

Figure 2.   Conventional submersible gate pump.

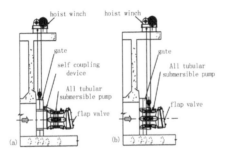

Figure 3.   All tubular submersible gate pumps.

application conditions, the installation type is of a self-coupling and fixed integral installation (Xu *et al.*, 2012). The self-coupling installation is similar to the conventional submersible gate pump. Particularly, when the water needs to be raised, the self-coupling hook of the submersible pump is inserted into the sliding track of the gate, and the submersible pump slides down along the track to the hook position on the gate to achieve the self-coupling of the submersible pump and the gate. However, the gate and the submersible pump should be installed separately, and the gate opening and closing, and the submersible pump lifting are also required to operate separately; therefore, the use and management are not very convenient. Fixed integral installation is to install the full-tubular submersible pump system (with a flap valve) on the gate by using a fixed flange, so that the pump and the gate are integrated into a whole. The barycenter of the whole system, which deviates from the gate, is smaller, so that there is a little impact on the gate structure. The pump can follow the gate to open and close at the same time, and lifting becomes more convenient. It is a conventional connection mode. The specific layout is shown in Figure 3.

3 CALCULATION OF THE GATE STRUCTURE STRESS AND THE OPENING/CLOSING CAPACITY

3.1 *The structural design of the inlet gate*

The inlet gate is a flat slider bearing structure composed of panels, beams, bearing steel tube, connecting flange ring, etc. The load distribution of the gate body is complex. In addition to the gate and submersible pump dead weight, there are static water pressure and torque of rotating flow on the gate when the pump is running, where the load direction is not in a plane, and the new open intake in the gate greatly weakens the gate structure (Yuan *L. G.*, 2012). Based on the above reasons, if the structure calculation is simplified as a plane problem, the conventional plane theory and the actual stress condition of the gate are largely different. In order to simulate the actual application condition, the 3D linear elastic finite element method is used to analyze the stress and deformation of the gate. The 3D analysis model is built based on the 3D *SolidWorks* design platform, and the *SimulationXpress* finite element analysis software (Wang *et al.*, 2009, Wang *et al.*, 2011 and Chen *et al.*, 2013) is introduced to calculate and analyze the stress and deformation displacement and other mechanical properties, to optimize the structure and to determine the appropriate safety factors, striving to match the actual conditions.

1. The finite element analysis theory

The basic principle of the finite element method is to divide the model into a finite number of nodes and elements, and the equations of the whole field variables are finally set up in the static strength and stiffness analysis by integrating all the elements of the interpolation function. In each unit, the interpolation function is used to represent the field variable, and the interpolation function is determined by the node value. The physical solution equations are set up through the node between the elements, and then by solving the equations, the analysis results are obtained:

$$[K]\{\delta\} = \{\mathbf{P}\}, \tag{1}$$

where [K] = the total stiffness matrix; $\{\delta\}$ = the displacement matrix; and {P} = the force matrix on each node.

2. Finite element analysis by *SimulationXpress*

SimulationXpress is the design analysis software based on the finite element analysis, fully integrated in the 3D design software of *SolidWorks*. It is able to provide pressure, constraints, frequency, and optimization design analysis, and in the *SolidWorks* environment, it is able to provide a more complete analysis means. With advanced Fast Finite Element Analysis (FFEA), it can function very rapidly to realize the analysis and verification of a large-scale complex design, and to obtain the necessary information to correct and optimize the design.

Meanwhile, the model and the results of the analysis will share the same database with *SolidWorks*. Moreover, the calculation results can be directly displayed on the *SolidWorks* precise design model, more intuitive to manage task analyses, loads, boundary conditions, finite element grid, and result data.

3. Design and establishment of a model

According to the 2D CAD inlet gate figure drawing to determine the size of the section of the plates, beams, steel pipes and flanges, and other components, the corresponding sketch is drawn in the *SolidWorks* 3D solid modeling environment. The 3D entity model of the inlet gate can be established by the drawing commands of matrix feature modeling. In the modeling process, we must pay attention to the elimination treatment of the model details. It should be based on the gate structure and working condition analysis, in order to choose some non-critical features such as "chamfering", etc. By importing the 3D entity model into the *SimulationXpress* environment, it can be automatically transformed into an effective finite element model and can be successfully carried out by the finite element mesh generation and calculation. The 3D finite element model of the gate is shown in Figure 4.

4. Define the material, impose constraints, and mesh division

Define the material: in *SimulationXpress*, an example of static numerical calculation is created, which is named the inlet gate structure analysis. First, the nature of the gate material is given, and the indicators of the material are defined, as given in Table 1.

Impose constraints and loads: gate body is installed in the gate slot, the side column on both sides of the gate is each set up with four wedge bearing, whose size is 150 × 400 mm. They are described as four end faces in the constraints of the door body and each end surface bearing has a constraint in the X, Y, and Z directions (green arrow), while the panel bottom has constraints in the Y direction (green arrow). The load of the gate has the weight of the gate (to be automatic), hydrostatic pressure unit weight (horizontal red arrow), the weight of the pump (purple arrow in the vertical direction), bolt tension connection flange (horizontal purple arrow), and torque on flange after the pump running (yellow arrow in the direction of rotation on the surface of the flange). The specific layout is shown in Figure 4.

Cell division: *SimulationXpress* can automatically divide the grid shape and size according to the gate, and it can also adjust itself to the grid size according to the need. The smaller the cell is and the more subtle the division of the grid is, the more cells there will be so that the

Figure 4.   Gate load model.

Table 1.   The indicators of the gate materials.

Material	Modulus of elasticity (N/m²)	Poisson's ratio	Shear strength (N/m²)	Mass density (N/m³)	Yield strength (N/m²)	Ultimate strength (N/m²)
Q235B	2e + 11	0.28	7.9e + 10	7.8e + 04	2.35e + 08	4.0e + 08

corresponding calculation will be bigger. In general, the results and accuracy can meet the requirements of the analysis by using the default grid division. Commonly used cells are three pyramid tetrahedron cells. The cell size is 55.259 mm and tolerance is 2.7630 mm. The gate is divided into 73,194 cells and the node number is 145,304. The finite element mesh diagram is shown in Figure 5.

5. Finite element analysis of stress and strain

When the parameter design is completed, *SimulationXpress* will automatically analyze and calculate the stress and deformation displacement nephogram. Finally, the results of the finite element analysis are compared with the calculation and analysis of plane theory to provide a reliable theoretical basis for determining the size of the overall structure and to optimize the structure. Under the premise of safety, it should be determined whether the size and material saving should be reduced for the finite element optimization design.

6. Optimization design

Optimization design is a kind of technology that searches for the optimal design scheme that can satisfy all the design requirements and the minimum expenditure, such as weight, area, volume, stress, fees, etc. In this paper, according to statics modal analysis results of the inlet gate, the safety factor is uniform in the condition of satisfying the structure size and stress limit. In a process of optimization design, a set of basic parameters considered as design variables are adjusted, in order to determine the objective function of the design and to control the value of the design variables in accordance with the actual requirements.

Optimization analysis of shape, size, and stress and strain characteristics: the inlet gate is a space thin-walled structure. The beam grid and the plate are welded to form a whole, so that the structural characteristics of the plate and the beam are not obvious. If there is a large difference between the stress and strain, it will not only waste the materials but also easily form a fatigue stress failure of the stress peak area under the dynamic water load of the submersible pump. Therefore, the stress and strain uniformity is established as the objective function in the gate optimization design. The gate, steel pipe, flange thickness, and the height of the beam system are considered as design variables, and the appropriate constraints are set according to the requirements of the specific situation. Based on the above analysis, optimization of the 3D entity model is established, and the reasonable stress and strain safety factors are obtained.

Optimization measures and security check: according to the results of the analysis, the stress and strain tend to be uniform by adjusting the thickness of the plate, the height of the beam, and the position and the number of the rib plate, and the stress safety factor (ratio of material yield limit to calculated stress) is controlled in the range of 3.0–15, and the strain coefficient (ratio of maximum strain and minimum strain) is controlled in the range of 5–10.

3.2 *Design of connecting flange and bolt structure*

The gate is connected to a full-tubular submersible pump with high strength bolts and flange ring. Although there are preliminary results in the 3D space finite element calculation, the

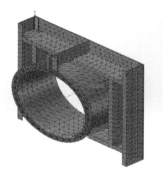

Figure 5.   Gate finite element meshes.

friction force and the bolt tightening pretension are unable to react in the finite element calculation, because they belong to the internal forces in the gate's structure calculation. Thus, the plane theory is needed to carry out the stress check to the bolts and the flange.

1. Resultant force of the connecting bolt ΣP is given by:

$$\Sigma P = P_{\Sigma_A} + P_g + P_m + P_X \tag{2}$$

Here $P_{\Sigma_A} = \pi D \delta \sigma_{X3}$, $P_g = KPF_g$

where P_{Σ_A} is the resultant force caused by the axial force and the friction force in kN; P_g is the tightening bolt pre-tension in kN; P_m is the force produced by the temperature in kN; P_x is the other axial force in kN; D is the diameter of the bearing steel tube in m; δ is the calculation thickness of the tube wall in mm; δ_{x3} is the pipe wall stress induced by the axial force in kN; K is the coefficient due to the water stop piece form; P is the design pressure of the supporting tube in kN/m²; and F_g is the area of the water sealing gasket in m².

2. Tensile stress of connecting bolts σ is given by:

$$\sigma = \sum P / A_n, \tag{3}$$

where ΣP is the resultant force of the connecting bolt in kN and A_n is the net area of the bolt in m².

3. Structural calculation of the flange ring

All tubular submersible pumps and gates are connected to an organic whole by bolts. Particularly, the pump is connected to the flange ring through a bearing steel tube of the gate, and the supporting steel pipe is welded together with the flange ring, which can be considered as a free flange with a neck pipe (Tie H., 2006).

$$\begin{cases} \sigma_{Z(B)} = \pm \dfrac{M^3}{r_2 \delta_f^2}(\varphi_3 - \dfrac{\varphi_4}{1+\dfrac{\delta_f^3 \varphi^2}{r_2 \delta^5}}) \\[4mm] \sigma_{Y(B)} = \pm \dfrac{M^3}{r_2 \delta_f^2}(\varphi_3 - \dfrac{\varphi_1}{1+\dfrac{\delta_f^3 \varphi_2}{r_2 \delta^5}}), \\[4mm] \sigma_{X(C)} = \sigma_{Y(B)}(\dfrac{\delta_f}{\delta})^2 \end{cases} \tag{4}$$

where δ_f is the thickness of the flange ring in mm; δ is the thickness of the neck tube in mm; h is the calculated height of the center of the flange to the end of the neck tube in m; r_1 is the outside radius of the flange ring in m; r_2 is the inside radius of the flange ring in m; r_3 is the circle radius of the bolt hole in m; r_4 is the radius of water seal center on the flange ring in m; and ψ_1, ψ_2, and ψ_3 are the calculation function, according to the related curve chart (Tie H., 2006).

3.3 *Outlet flap valve structure design*

The outlet flap valve is arranged away from the gate and lifting position. In order to reduce the impact of gate bias caused by the weight of the flap valve to ensure the suitable opening angle of the flap valve, the flap valve materials and open direction are optimized. The cast iron material, which easily rusts and is heavy, is discarded, and the HDPE non-metal material that is lightweight and corrosion resistance is selected (Cui *et al.*, 2013, Zhang *et al.*, 2005). The way of opening the gate is optimized, and the energy-saving and free side open flap valve is used. The opening angle is 2/3 larger than the traditional hanging flap valve, so the flow velocity and the flow rate are increased by about 30%, and the plane efficiency of the pump station is thus improved (Yang *et al.*, 2011).

4 THE ENGINEERING EXAMPLES

Hongye drainage culvert is located in Yizheng City in Jiangsu Province of China, which is the main drainage channel in Hongye district of Yizheng City. There is only one opening culvert in total with the size of 2.5 × 2.5 m, cast iron gate is set in the culvert mouth, controlled by QL-80 kN-SD screw hoist. The drainage culvert was built in 2008, and its design discharge is 8 m³/s.

By completion, the Hongye drainage culvert can meet the design requirements of the gravity flow drainage. However, when it rains heavily and the outer river is in high water level, the inland river flood cannot be discharged so that the regional flooding will happen frequently. What is worse is the living sewage of coastal residents is directly discharged into the river, and the river water cannot be exchanged in time. As a result, there is an urgent need to increase the drainage facilities due to the serious eutrophication of the water quality.

4.1 Structure arrangement of the gate pump and the pump type selection

In view of this project, the pumping flow rate is small (the design discharge Q = 3.0 m³/s), and the head (the design head Hs = 2.5 m) is low, the pump type of the full-tubular submersible pump is adopted. The welding slider steel gate is used as the retaining water structure. By the use of the new gate, the inlet hole is made on the panel. The outlet of the submersible pump is connected to the flap valve as a shut off equipment. The supporting steel pipe is welded and connected with the connecting flange ring at the open hole position and is fixedly connected with the full-tubular submersible pump with the high strength bolts. When gravity draining and maintenance, the gate pump can be lifted over the water surface as a whole, quickly transforming function. If the river is in high water level and holding back water or pumping is needed, gate pump can be dropped to the bottom of the floor as a whole so that the engineering requirements of gravity draining, water exchange, holding back water, and pumping can be met.

According to the design flow and head, and related technical parameters, a type of 1000QGWZ-135 J full-tubular submersible pump is installed. The overall length of the pump is 1.47 m, rotational speed is n = 423 r/min, and its impeller diameter is 950 mm. The main technical parameters of the pump are shown in Table 2. Its contour structure size and performance curves are shown in Figures 6 and 7, respectively.

4.2 Stress calculation of the gate structure

1. The inlet gate structure calculation
The inlet gate is a wedge slider bearing plate structure, which is composed of the panel, beam, connecting flange ring, etc. The sluice opening width is 2.5 m, and the gate leaf size is 2.93 × 2.90 × 0.40 m (width × height × thick), thus the safety factor of the dynamic water load is 1.5–2.0, and the diameter of the inlet hole is $\phi1.5$ m. Using the theory of 3D space finite element and the *SolidWorks* design platform and the *SimulationXpress* finite element analysis software, the structural analyses are carried out. Applied loads are listed in Table 3. The constraints are imposed for setting four slider supports, and the X, Y, and Z direction constraints assumed on the surface of the supporting slider, and the Y direction of the constraint on the lower side of the panel. According to the requirements of the finite element optimization

Table 2. 1000QGWZ-135 J all tubular submersible pump performance parameters.

Traffic (l/s)	Pump head (m)	Leaf blade angle (°)	Rated speed (r/min)	Shaft power (kW)	The pump efficiency (%)	Match the motor	Impeller diameter (mm)
2916	3.00	−2	423	100.7	85.25	YQSN1180 −14	Φ950
3102	2.50			89.4	85.27	(132 kW, 380V)	
3428	1.58			65.8	80.62		

design, the optimization of structure of the inlet gate is executed, and the appropriate safety coefficient is selected. The nephogram of the optimized gate stress and deformation displacement are shown in Figures 8 and 9.

The calculation shows that the maximum Mises (an equivalent stress in the fourth strength theory, von Mises stress) stress is 63.4 MPa in the upper and the lower main beams, the maximum Ures (resultant force displacement) displacement is 0.55 mm. The supporting steel pipe and the flange have a larger displacement, basically in the range of 0.40–0.55 mm, and the maximal displacement of main beams is 0.45 mm. In addition to the main beams, the remaining stress is smaller, about 15 MPa. The stress and the strain are more uniform, and the control range of the stress safety factor is reasonable. The safety factor of the panel, beam, teel pipes, flanges, and other major components is controlled at 3.5. According to the requirements of the engineering construction, the maximum safety factor of stiffening plate

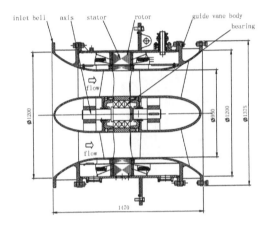

Figure 6.   Structure size of the gate pump.

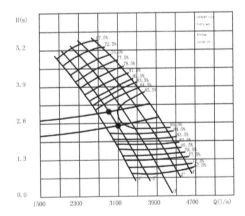

Figure 7.   Tubular submersible pump performance curve.

Table 3.   Load and constraints of the inlet gate.

Static water pressure (kN/m²)	Weight of the submersible pump (kN)	The pull of the bolt (kN)	Torque on the flange (kN.m)
218	45	5.5	7.7

25

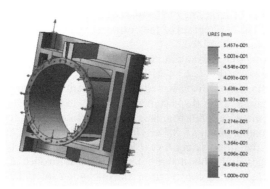

Figure 8.   Gate Ures displacement.

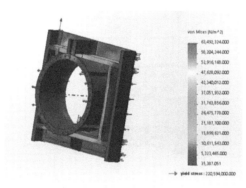

Figure 9.   Gate Mises stress.

and secondary components is 15. Therefore, the overall structure has a certain safety margin and the requirements of the optimization design can be confirmed.

2. Calculation of the flange and the connecting bolts

The full-tubular submersible pump is connected to the gate with bolts, where the flange ring of the supporting steel pipe is welded on the gate body to connect the pump. Because the supporting steel pipe is welded with the flange ring, the flange can be considered as a free one with a neck pipe. The thickness of the flange ring is 36 mm, the inside diameter of the flange ring is 1.5 m, the outside diameter is 1.75 m, and the connecting bolt is 36-M27. According to the plane calculation theory for the connection bolt and the flange strength review, the specific calculation results are listed in Tables 4 and 5, respectively.

According to the plane theory, the maximum tensile stress of the bolt (16.7 MPa) is close to that of the Mises stress, and the maximum stress of the flange (57.7 MPa) is larger than that of the maximum Mises stress (25 MPa) of the 3D finite element. When the flange is considered as a whole, the flange is strengthened by optimizing the number and locations of the plate that is stiffened with the diverted steel pipe, reducing the stress of flange neck. At the same time, it is necessary to optimize the design by *SimulationXpress* finite element analysis software.

3. The calculation of the outlet flap valve

In the outlet, a circular energy-saving, rollover-type, and a **HDPE** non-metallic flap valve is applied to cut off water. To ensure safety, the conventional suspension flap valve structure is adopted for the calculation condition of the opening angle of the flap valve and is verified through an experiment. The results are shown in Table 6.

Table 4. Connecting the bolt calculation results.

$P_{\Sigma A}$ (kN)	P_g (kN)	P_m (kN)	P_x (kN)	ΣP (kN)	A_n (m²)	σ_{max} (MPa)
218	45	5.5	7.7	276.2	1.65×10^{-2}	16.7

$P_{\Sigma A}$, P_g, P_m, P_x, A_n, and σ_{max} represent the resultant force caused by the axial force and the friction force, tightening bolt pre-tension, force produced by temperature, other axial force, resultant force of the connecting bolt, net area of the bolt, and the tensile stress of the connecting bolts.

Table 5. Connecting bolt calculation results.

δ_f (mm)	r_1 (m)	r_2 (m)	M_{max} (MPa)	$\sigma_{z(B)}$ (MPa)	$\sigma_{y(B)}$ (MPa)	$\sigma_{x(C)}$ (MPa)
36	0.875	0.756	10.2	9.6	11.2	100.3

δ_f, r_1, r_2, M_{max}, $\sigma_{Z(B)}$, $\sigma_{y(B)}$, and $\sigma_{x(c)}$ represent the thickness of the flange ring, outside diameter of the flange ring, inside diameter of the flange ring, the bending moment, the point B stress, the point B stress, and the point C stress.

Table 6. Calculation results of the flap valve opening angle.

α_B (°)	Q (m³/s)	V (m/s)	G (kN)	W (kN)	L_g (m)	α (°)
0	3.0	1.0	9.0	3.3	0.15	38.1

α_B, Q, V, G, W, L_g, and α represent the center of the pipe and the horizontal plane angle, the flow of the water pump, the flow rate of the pipe exports, the weight of the flap valve, buoyancy of the flap valve, the gravity of the flap valve to the center of rotation, and the opening angle of the flap valve.

According to the calculation results, the opening angle of a single fan hanging conventional flap valve is 38.1°. Through the experiment, the opening angle of energy-saving and rollover-type flap valve is about 58°, so that it can fully meet the engineering requirements

4.3 The use of the submersible gate pump

In low-head irrigation and drainage areas, such as the polder areas along the Yangtze river, low-lying areas along the Huai River, and Huaibei plain, full-tubular submersible gate pumps have a broad application prospect. In regions along the Yangtze river and the Huai River in Jiangsu, most of the places were built with an independent automatic drainage (or abstraction) culvert and sluice. The gravity flow drainage culvert and the sluice gate are rebuilt into gate pumps, as a supplement of the existing drainage pumping station, which can effectively improve the drainage capability. At present, Jiangsu Province has built gate pumps like Hongye gravity drainage culvert and Xupu River Sluice in Yizheng, Rugao River drainage culvert in Nantong, etc. Hongye gravity drainage culvert and Rugao River drainage culvert are located in the riverside of the Yangtze river, and they are both inland drainage culvert or sluice. When the Yangtze river flood jacks, in order to ensure the river flood is pumped and drained into the Yangtze river, they are converted into submersible gate pumps with the trash facilities set-up in the river side near the gate. Hongye gate pump was put into operation in February 2014 and Rugao River gate pump has been in trail run since March 2015. Both of them play a significant role in gravity drainage, water exchanging, pumping, etc. Under normal conditions, the inner and the outer river water body can be freely exchanged. When the outer river floods, the gate pump can be dropped down to block flood or pump drainage, so the flood control safety in that region can be guaranteed, and the surrounding environment can be improved.

The size of the sluice hole of the Xupu River check sluice is 4.5×4.0 m (width $\times$ height). The work gate is constructed at the side of the river, and the overhaul gate slot is set-up at the side of the Yangtze River. The original function of the sluices is timely diversion, drainage, and flood control. Due to the sluice being located in the city, where more and more living garbage causes inland water quality getting worse, it is often necessary to pump water to exchange water and activate water. The elevation of the sluice floor is relatively high, so the water diversion could be difficult when the water level of the Yangtze River is low. Using the way of the tubular submersible pump and the gate coupling arrangement, the tubular submersible pump is arranged on the stilling basin base plate of the Yangtze river side and the inlet channels of the steel structure are arranged in the inlet water side. By lowering the inlet height, it could satisfy the demands of the inlet flow state even though the water level is low in Yangtze river. 1400QGWZ-125 type tubular submersible pumps are adopted. The outlet pipe is connected to the gate in an overhaul gate slot through a coupling. The flap valve is arranged to block water on the outlet side gate, and QP-2 $\times$ 100 kN type, winch hoist is set as well. The outlet side gate (with little flap valve) can be raised from the Sluice hole. When the Yangtze river water level is low, we should open the gate, put down the outlet side gate of the access door slot, and switch on the submersible pump to pump water from the Yangtze River to exchange and maintain the river water level and the landscape water level at the same time. The function of the flap valve is to block water in shutdown. It has already ensured water transfer security for 81 days in the Yangtze river during dry season. Overall, it has acquired ideal results at no additional investment cost of civil construction.

5 CONCLUSION

1. To full-tubular submersible gate pump, a fixed integral connection is adopted for the gate and the pump, so pumps and gates are truly integrated. The center deviation of the whole device is minor, and it can open and close at the same time with the gate. Hence, the management is convenient. Also, it can be a good substitute for the small- and the medium-sized submersible axial flow gate pump and the semi-tubular submersible pump.
2. Because the load situation is complex for the inlet gate of the gate pump, the conventional plane stress method is not reliable. But by adopting the 3D space finite element theory and relying on *SolidWorks* 3D design platform and the *SimulationXpress* software for structural optimization design and analysis, we can well solve the structural problems under complex conditions, ensuring a safe and a reliable operation of the project.
3. In the polder areas along the Yangtze River, low-lying areas along the Huai River, Huaibei Plain, and other low-lift irrigation and drainage areas, the way of rebuilding discharge culvert and checking sluice into gate pump is feasible without an additional increase in the work quantities of Civil Engineering. Drainage capacity is promoted due to reconstructing the gate pump from the drainage culvert. By using the maintenance gate in the regulator gate to rebuild the gate pump and interactively running both the gate pump and the working gate, the water turning, the drought resistance capacity, and the water quality are improved. The arrangement of "one gate one pump" or "one gate two pump" increases the diversity of the project layout. The functions of drought resisting, flood drainage, and water exchange are further enlarged, which significantly improves the efficiency of the project and saves the investment of the project.

ACKNOWLEDGMENTS

This paper was supported by the 12th Five Year Key Project of China's National Scientific Supporting Plan (No. 2015BAB07B01).

REFERENCES

Chen C.Q & Hu Q.D., 2013. *SolidWorks*-Simulation advanced tutorial. *Beijing: China Machine Press*.

Cui Y.S., Ma J. & Li D.W., et al., 2013. Specification for design of steel gate in hydraulic and hydroelectric engineering. *Beijing: China WaterPower Press*.

Gu Z.S., 2010. The development and application of tubular submersible pumps. *Water Resources and Hydropower Engineering*, 12(6):36–38.

Jing L., 2015. The development of built-in submersible gate pump impeller. *Pump Technology*, 10(2):23–26.

Li L., Wang Z. & Hu R.X, et al., 2007. Numerical analysis of hydraulic performance for the bidirectional tubular pump. *Transactions of the Chinese Society for Agricultural Machinery*, 38(1):76–79.

Liu C., 2015. Researches and Developments of Axial -flow Pump System.*Transactions of the Chinese Society for Agricultural Machinery*, 46(6):49–59.

Rong Z.F., 2004. Energy-efficiency measures for small and medium-sized axial-flow pump station. *Drainage and Irrigation Machinery*, 12(1):20–23.

Tie H., 2006. Analysis and stress calculation of pressure steel pipe into the manhole boundary effect. *Xibai Hydro & Power*, 4(1):11–14.

Wang S.L., 2009. Submersible gate pump development and successful application. *Pump Technology*, 6(2):54–57.

Wang W.B., Tu H.L. & Xiong J.X., 2009. *SolidWorks*2008 development basis and examples. *Beijing: Tsinghua University Press*.

Wang Y. & Ye Q., 2011. Finite element analysis and optimization design of connecting rod based on *SolidWorks. J. Zhejiang Wat. Cons & Hydr. College*, 23(4):51–53.

Wu C.X., Jin Y., Wang D.H., et al.,2012. The research and application of large bidirectional through-flow submersible pump device, *Journal of Hydroelectric Engineering*, 18(6):24–26.

Xu J.Z., Gu Z.S. & Lv J.S., et al.,2012. Technical Code for Submersible Pumping Station (SL584–2012). *Beijing: China WaterPower Press*.

Yang F., Zhou J.R. & Liu C., 2011. Numerical simulation and experiment on resistance loss of flap gate. *Transactions of the Chinese Society for Agricultural Machinery*, 42(9):108–112.

Yuan L.G., 2012. The application of wedge block and the ramp in the mobile pumping station pumps brake combined bearing seat and the steel gate. *Hunan Hydro & Power*, 3(2):27–31.

Zhang D.S., Shi W.D. & Chen B., et al. 2010. Unsteady flow analysis and experimental investigation of axial-flow pump. *Journal of hydrodynamics*, 22(1):35–43.

Zhang J.P., Li D.H. & Sun W.X., et al., 2005. The study of no modern metal gate. *Journal of Hydraulic Engineering*, 28(12):25–26.

Hydraulic Engineering IV – Xie (Ed.)
© *2016 Taylor & Francis Group, London, ISBN 978-1-138-02948-4*

New explicit analytical solutions of rogue wave based on the HAM

L. Zou, Z.B. Yu & Y.G. Pei
*State Key Laboratory of Structural Analysis for Industrial Equipment, School of Naval Architecture,
Dalian University of Technology, Dalian, China*
Collaborative Innovation Center for Advanced Ship and Deep-Sea Exploration, Shanghai, China

Z. Wang
School of Mathematical Sciences, Dalian University of Technology, Dalian, China

ABSTRACT: Rouge waves are very large, rare events in a random ocean wave train. In this paper, the homotopy analysis method is applied to obtain Nth-order rogue wave solution of nonlinear Schrödinger equation. In particular, the interesting structures for the fifth- and seventh-order solutions of rogue wave are shown. The analytical results will guide the experiments and can help generate rouge wave in the laboratory in the future.

1 INTRODUCTION

Rogue waves are spontaneous nonlinear waves with amplitudes significantly two times higher than the surrounding average wave crests and have potentially devastating effects on offshore structures and ships. For a quite long time, rogue waves are thought to be mysterious, as they appear from nowhere and disappear without a trace. In real circumstances, many physical mechanisms are known to contribute to this phenomenon: dispersion enhancement, geometrical focusing, wave–wind and wave–current interactions, and modulational instability. Of course, it is hardly possible to include all of those ingredients into a single theoretical model. Extreme waves that may be well represented by the Peregrine soliton have recently been experimentally observed in optical fibers (Kibler, B. et al. 2010) and in water wave tanks (Chabchoub, A. et al. 2011). Zha (Zha, Q. L. 2013a) investigated the Generalized Nonlinear Schrödinger (GNLS) equation by Darboux matrix method. Efe (Efe, S. et al. 2015) studied the discrete rogue waves in an array of wave guides. Priya (Priya, N. V. et al. 2015b) constructed a Generalized Darboux Transformation (GDT) of a General Coupled Nonlinear Schrödinger (GCNLS) system. However, all these studies are based on the exact solutions of rouge waves (Zhang, Y. et al. 2014). In this paper, we first present the rogue wave solution of the nonlinear Schrödinger (NLS) model based on the asymptotical theory. We then successfully obtain the explicit analytical nth-order approximations of rogue waves. Our analytical results will guide the experiments and can help generate rouge wave in the laboratory in the future.

2 MATHEMATICAL FORMULATION

Here, we consider the NLS equation (Zha, Q, L. 2013b):

$$i\left(\psi_t + C_g \psi_x\right) + \mu\psi_{xx} + \upsilon|\psi|^2\psi = 0 \tag{1}$$

where $C_g = \dfrac{1}{2}\dfrac{\omega_0}{k_0} = \dfrac{1}{2}\dfrac{L_0}{T_0}$, $\mu = -\dfrac{\omega_0}{8k_0^2}$, $\upsilon = -\dfrac{1}{2}\omega_0 k_0^2$, and $\lambda = \sqrt{\dfrac{\upsilon}{2\mu}} = \sqrt{2}k_0^2$. Here the subscript "0" denotes the carrier wave that is modulated by the function $\psi(x,t)$: k_0 is the wavenumber, ω_0

is the carrier wave frequency, L_0 is the carrier wave length, and T_0 denotes the carrier wave period. The aim of this paper is to obtain rogue wave solution of the NLS equation. The main tool is a simple symbolic computation approach. The complex field $\psi(x,t)$, as appearing in the NLS equation, is related to the amplitude and phase of the water waves. Consider a slowly modulated carrier wave with water surface elevation $\eta(x,t)$ of the form:

$$\eta(x,t) = Re\left[\psi(x,t)e^{i(k_0x-\omega_0t)}\right]/2 \tag{2}$$

Thus, the carrier $e^{i(k_0x-\omega_0t)}$ is modulated by the complex envelope, $\psi(x,t)$ as determined in equation [1] for some chosen initial condition, $\psi(x,0)$. The simple transformation:

$$u = \lambda\psi, T = \mu t, X = x - C_gt \tag{3}$$

allows equation [1] to be put into non-dimensional form. The non-dimensional NLS equation arises:

$$iu_T + u_{XX} + 2|u|^2 u = 0 \tag{4}$$

where λ is a scale factor. This simple form of NLS is often used for mathematical convenience.

3   RESULTS AND ANALYSIS

Consider the NLS equation in the non-dimensional form of

$$iu_T + u_{XX} + 2|u|^2 u = 0 \tag{5}$$

which is subjected to the initial condition:

$$u(X,0) = g(X) \tag{6}$$

In the complex plane, we obtain the following equation:

$$|u|^2 = u\bar{u} \tag{7}$$

Multiplying i both sides of equation [5], we obtain the following equivalent equation:

$$u_T - iu_{XX} - 2i|u|^2 u = 0 \tag{8}$$

According to equation [5], we define a nonlinear operator as follows:

$$\mathcal{N}[U(X,T;q)] = \frac{\partial U(X,T;q)}{\partial T} - i\frac{\partial^2 U(X,T;q)}{\partial T^2} - 2iU(X,T;q)^2 \overline{U(X,T;q)} \tag{9}$$

Then, we choose the initial approximation as follows:

$$u_0(X,T) = \text{sech}(X) \tag{10}$$

and the auxiliary linear operator as:

$$\mathcal{L}[U(X,T;q)] = \frac{\partial U(X,T;q)}{\partial T} \tag{11}$$

32

It possesses the following property:

$$\mathcal{L}[C_1] = 0 \tag{12}$$

where C_1 is coefficient. Similarly, we construct the zeroth-order deformation equation (Liao, S. J. 2003, Tan, Y. 2007) as follows:

$$(1-q)\mathcal{L}[U(X,T;q) - U_0(X,T)] = qh\mathcal{N}[U(X,T;q)] \tag{13}$$

which is subjected to the initial condition:

$$U(X,0;q) = \operatorname{sech}(X) \tag{14}$$

When $q = 0$ and $q = 1$, we obtain from the zeroth-order deformation equation [13] that $U(X,T;0) = u_0(X,T)$ and $U(X,T;1) = u(X,T)$. Especially, we define $u_m(X,T)$ in the following form:

$$\overline{u_m(X,T)} = \frac{1}{m!} \frac{\partial^m U(X,T;q)}{\partial q^m}\bigg|_{q=0} \tag{15}$$

In the similar way, we obtain the mth-order deformation equation as follows:

$$\mathcal{L}[u_m(X,T) - \chi_m u_{m-1}(X,T)] = hR_m(\vec{u}_{m-1}) \tag{16}$$

which is subjected to the initial condition:

$$u_m(X,0) = 0, (m \geq 1) \tag{17}$$

where

$$R_m(\vec{u}_{m-1}) = \frac{\partial u_{m-1}}{\partial T} - i\frac{\partial^2 u_{m-1}}{\partial X^2} - 2i\sum_{n=0}^{m-1} \vec{u}_{m-1-n} \sum_{j=0}^{n} u_j u_{n-j} \tag{18}$$

$$\chi_m = \begin{cases} 0 & m \leq 1 \\ 1 & m > 1 \end{cases} \tag{19}$$

It is easy to get the solution of the above linear differential equation. Its solution $u_m(X,T)$ can be expressed in the form:

$$u_m(X,T) = \chi_m u_{m-1} + h\int_0^T R_m(\vec{u}_{m-1})dT + C_1 \tag{20}$$

where C_1 is an integral constant to be determined by initial condition equation [17].

Then, the solution to the equation [5] can be written in an accurate form:

$$u(X,T) = u_0(X,T) + \sum_{m=1}^{+\infty} u_m(X,T) \tag{21}$$

The condition is given by:

$$u(X,0) = \operatorname{sech}(X) \tag{22}$$

and the initial approximation is chosen as:

$$u_0(X,T) = \operatorname{sech}(X) \tag{23}$$

33

Substituting it into the high-order deformation equation [16], it is easy to get the solution to the linear differential equation, especially by means of symbolic software such as Mathematica, Maple, and so on. Its solution can be expressed in the form when $h = -1$:

$$u_1(X,T) = iT\text{sech}(X)$$
$$u_2(X,T) = -\frac{1}{2}T^2\text{sech}(X)$$
$$u_3(X,T) = -\frac{1}{6}iT^3\text{sech}(X)$$
$$\vdots$$

(24)

Similarly, we also can obtain the exact explicit solution:

$$u(X,T) = \sum_{k=0}^{+\infty} u_k(X,T)$$
$$= \sum_{n=0}^{+\infty} \frac{(iT)^n}{n!}\text{sech}(X)$$
$$= e^{iT}\text{sech}(X)$$

(25)

A few typical examples are displayed in Figures 1, 2, 3, and 4 by two cases. We present that the profile of rogue wave for the wave number k is 1. The profiles of the rogue waves are illustrated in Figures 1 and 2.

Set $A = \psi(x,t)e^{i(k_0x - a_0t)}$, by asymptotical, computation, and when $k = 1$, we get the following equation:

$$A = \frac{1 - 0.39it - 0.076t^2 + 0.01it^3}{2\cosh(x - 1.56t)}\sqrt{2}e^{i(x - 3.13t)}$$

(26)

Set $A = \psi(x,t)e^{i(k_0x - a_0t)}$, by asymptotical computation, when $k = 2$, we get the following equation:

$$A = \frac{1 - 0.13it - 0.01t^2 + 0.0004it^3}{8\cosh(x - 1.1t)}\sqrt{2}e^{i(2x - 4.4t)}$$

(27)

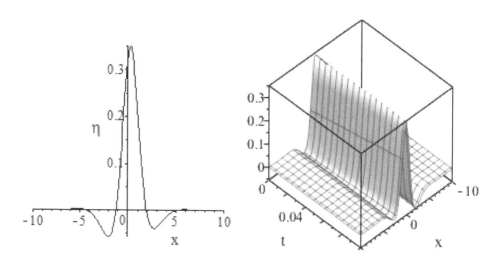

Figure 1.   The fifth-order solution at the real part of $(A/2)$ when $k = 1$ and $t = 0.1$.

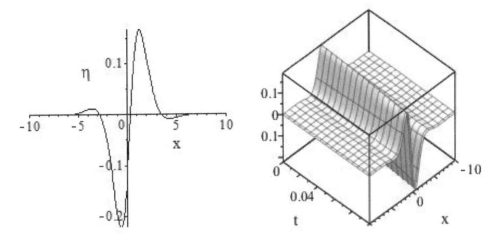

Figure 2.   The fifth-order solution of the imaginary part of $(A/2)$ when $k=1$ and $t=0.1$.

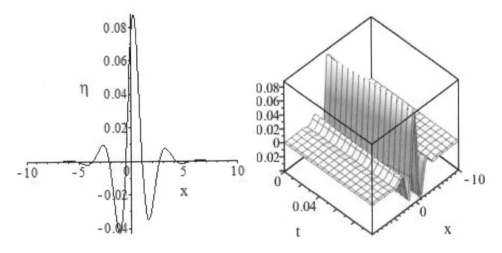

Figure 3.   The seventh-order solution at the real part of $(A/2)$ when $k=2$ and $t=0.1$.

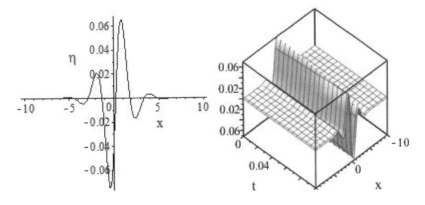

Figure 4.   The seventh-order solution of imaginary part of $(A/2)$ when $k=2$ and $t=0.1$.

Figures 3 and 4 show the wave surface profiles of rogue waves. The nth-order approximation is obtained by the homotopy analysis method. We hope that our analytical results will be realized by physical experiments in the future, which can be used to understand the generation mechanism and to find possible applications of the rogue wave for the NLS equation (Borhanian, J. 2015).

4 CONCLUSION

An analytical study of the properties of rogue waves was performed based on explicit analytic solutions of the NLS equation that described the nonlinear evolution of deep water wave trains. The results of this paper can be used to understand the generation mechanism and to find possible applications of the rogue wave (He, j. S. et al. 2014, Priya, et al. 2015a). Our analytical results will guide the experiments and can help generate rouge wave in the laboratory in the future.

ACKNOWLEDGMENTS

This work was supported by 973 Program (2013CB036101), the National Natural Science Foundation of China (51379033, 51522902 and 51579040), and the Fundamental Research Funds for the Central Universities (DUT15 LK43 and DUT15 LK30).

REFERENCES

Borhanian, J. 2015. Extraordinary electromagnetic localized structures in plasmas: Modulational instability, envelope solitons, and rogue waves. *Phys. Lett. A*, **379**(6): 595–602.
Chabchoub, A. et al. 2011. Rogue Wave Observation in a Water Wave Tank. *Phys. Rev. Lett.*, **106**:204502.
Efe, S. & Yuce. C. 2015. Discrete Rogue waves in an array of waveguides. *Phys. Lett. A*, **379**(18–19): 1251–1255.
He, J.S. et al. 2014. Rogue waves in nonlinear Schrödinger models with variable coefficients: Application to Bose-Einstein condensates. *Phys. Lett. A*, **378**(5–6): 577–583.
Kibler, B. et al. 2010. The Peregrine soliton in nonlinear fibreoptics. *Nat. Phys.*, **6**(10): 790–795.
Liao, S.J. 2003. Beyond perturbation: Introduction to the Homotopy Analysis Method. London: Chapman & Hall/ CRC Press,
Priya, N.V. & Senthilvelan, M. 2015a. On the characterization of breather and rogue wave solutions and modulation instability of a coupled generalized nonlinear Schrödinger equations. *Wave Motion*, **54**: 125–133.
Priya, N.V. & Senthilvelan, M. 2015b. Generalized Darboux transformation and Nth order rogue wave solution of a general coupled nonlinear Schrödinger equations. *Commun. Nonlinear Sci. Numer. Simulat*, **20**(2): 401–420.
Tan, Y. 2007. Series solutions of boundary-layer flows with algebraically decaying property. *Phys. Lett. A*, **367**(4–5): 307–310.
Zha, Q.L. 2013a. On Nth-order rogue wave solution to the generalized nonlinear Schrödinger equation. *Phys. Lett. A*, **377**(12): 855–859.
Zha, Q.L. 2013b. Rogue waves and rational solutions of a (3+1)-dimensional nonlinear evolution equation. *Phys. Lett. A*, **377**(42): 3021.
Zhang, Y. et al. 2014. Rogue wave solutions for the coupled cubic–quintic nonlinear Schrödinger equations in nonlinear optics. *Phys. Lett. A*, **378**(3): 191–197.

Hydraulic Engineering IV – Xie (Ed.)
© 2016 Taylor & Francis Group, London, ISBN 978-1-138-02948-4

A monitoring model based on MIV-improved RBF neural network for dam deformation

Xinyang Ning & Xiaoqing Liu
College of Water Conservancy and Hydropower Engineering, Hohai University, Nanjing, China

ABSTRACT: Given that the dam monitoring model based on the conventional Radial Basis Function (RBF) neural network cannot screen the remarkable factors and is liable to fall into the local optimum, this paper establishes a fusion monitoring model that combines the Mean Impact Value (MIV), improved fruit Fly Optimization Algorithm (FOA), and the RBF neural network. First, the MIV is introduced to screen three kinds of forecast factors, namely, water pressure, temperature, and aging. Then, the improved FOA algorithm is adopted for searching the optimal *spread* value of the RBF network. By using these two methods, the MIV-improved RBF neural network model is set up. To verify the validity of the model, taking the displacement monitoring data of the gravity dam into account, the multiple linear regression model, the conventional RBF model, and the MIV-improved RBF model are also built. The calculation results show that the MIV-improved RBF neural network model has characteristics such as great generalization ability, stable prediction, and high precision. Furthermore, this model can be applied to the dam deformation for monitoring and warning.

1 INTRODUCTION

Research on dam deformation monitoring and prediction has an important influence on the safe operation of the dam. The linear regression model is usually used to analyze the data of displacement in the conventional dam deformation monitoring model (Gu, 1990). However, the actual dam deformation has a complex nonlinear problem, which is influenced by geology, topography, construction material, and other factors. Thus, the prediction results of the regression model often have a low accuracy and a bad imitative effect. Recently, an artificial neural network technology has been developed and applied to solve nonlinear and uncertainty problems, because it does not consider the specific physical meaning of the introduced variable, but focuses on the relationship between the amounts of data (Zhang, 2003).

Compared with the Back-Propagation (BP) neural network, the Radial Basis Function (RBF) neural network has a more simple structure and is quite competent in handling with strong nonlinear problems (Cao, 2010), and the RBF model has been widely applied to solve the nonlinear problem of safety monitoring. Zhang (2003) established the RBF neural network model of dam section deflection prediction. Huang (2014) used ant-based clustering algorithm to construct the basic function of hidden layer of the RBF neural network model. Lv (2012) introduced the POS algorithm in traditional RBF to improve the fitting effect of traditional RBF.

However, the screening of the forecasting factor is ignored in the conventional RBF neural network model, and the introduction of irrelevant factors will directly lead to unstable results, which will be difficult to explain. Meanwhile, the generalizability of a weak network will make the RBF network fall into a local optimal solution. The Mean Impact Value (MIV) is one of the best indicators for evaluating the correlation of variables in the neural network. At the same time, studies in the literature (Xu, 2015; Li, 2014) have indicated that the stability and the generalizability of fruit Fly Optimization Algorithm (FOA)–RBF is superior to Particle Swarm Optimization (PSO)–RBF and Genetic Algorithm (GA)–RBF, and the improved FOA method performs much better during the optimization.

In this paper, the MIV is introduced to carry out the screening of forecast factors, and an improved FOA algorithm is adopted to optimize the value of *spread* in the RBF neural network. Based on the MIV-improved RBF neural network, the deformation forecasting model of gravity dam is built for deformation monitoring data and its effectiveness is verified.

2 MIV ALGORITHM

MIV method is put forward by Dombi *et al.*, which can measure the influence degree of input neurons on the output neurons. The specific calculation process is as follows. First, the original sample M is trained by the neural network; after training the sample M, the value of input neurons of the variables in the sample M is, respectively, changed by $\pm 10\%$; as a result, input neurons of two new samples ($M1$ and $M2$) are produced; $M1$ and $M2$ are trained again by the neural network and the prediction results are $N1$ and $N2$, respectively. Second, the difference in value between $N1$ and $N2$ is calculated; moreover, to obtain the MIV of input neurons, this value is averaged by the number of input neurons. (The plus or the minus of the MIV represents the direction of relation, and the size of the absolute value represents the degree of relation.). Finally, As the MIV of input neurons is bigger, the impact of input neurons is greater on the output neurons, and the significant input neurons are screened to find out the significant predictor of the monitor model (Xu, 2012).

Through the above-mentioned process, it is easily found that the MIV algorithm is simple and can be applied to the complex nonlinear problem due to the training means of the neural network.

3 RBF NEURAL NETWORK MODEL

3.1 *RBF neural network model*

The RBF neural network model (three layers forward neural network), composed of the input layer, the hidden layer, and the output layer, is suitable for dealing with different kinds of nonlinear problems. The basic principle of the model is that when the space of hidden layers is built, based on the hidden RBF unit, the low-dimensional input neurons will be transformed to the high-dimensional space and decomposed. The structure diagram of the model topology is shown in Figure 1. Input neurons are transformed from the input space to the hidden space, using the Gaussian function (see Equation [1]) and the mapping relation function of RBF neural network (see Equation [2]) (Cao, 2010).

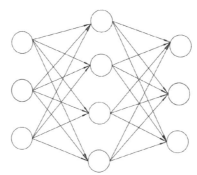

Input layer   Hidden layer   Output layer

Figure 1.   Diagram of the RBF network topology.

$$\varphi\left(\|x-c_p\|\right)=\exp\left(\frac{\|x-c_p\|}{2\sigma_p^2}\right) \tag{1}$$

$$g(x)=w_0+\sum_{i=1}^{p}w_i\varphi\left(\|x-c_p\|\right)=w_0+\sum_{i=1}^{p}w_i\exp\left(\frac{\|x-c_p\|}{2\sigma_p^2}\right), \tag{2}$$

where x is the input neurons, p is the node number of the hidden layers, w_i is the weight coefficient, c_p is the base function center, and σ_p is the width of the Gaussian function.

The RBF neural network can be established by the function of MATLAB, and the format is as follows:

$$net=newrbe(P,T,spread) \tag{3}$$

where P is the input neurons, T is the output neurons, and *spread* is the extending speed.

In fact, the value of *spread* directly affects the forecasting performance of the RBF neural network, that is to say, if the value is too big, it will lead to overlapping RBF neural network of neurons excessively; on the other hand, if the value is too small, the network cannot contain all the input area of the neuronal extension (Xu, 2015).

3.2 The improved FOA algorithm

FOA is a kind of intelligent optimization algorithm, based on the foraging behaviors of fruit flies (Pan, 2012). Owing to its strong intelligence, this algorithm has been broadly utilized in the function analysis, artificial intelligence, optimization of neural network model parameters, and so on (Xu, 2015). However, during the process of optimization, the unreasonable setting of the search radius r in the conventional FOA algorithm easily leads to the local minima or misses the optimal solution algorithm. Thus, it is necessary to adopt an improved FOA algorithm, which can avoid the limitations of the conventional FOA by adjusting the searching radius r corresponding to the different number of iterations (odd or even) (Li, 2014).

When the number of iterations is odd, let the algorithm search the global scope; in other words, set the value of the searching radius r large enough, and when the number of iterations is even, let the algorithm search near the optimal value; in other words, set the value of the searching radius small enough. The cycle is to ensure the algorithm to find the optimal solution in the global area, and the value r is calculated as follows (see Equations [4], [5], and [6]):

$$\Delta Smellbest=Smellbest_i-Smellbest_{i-1} \tag{4}$$

$$r=\frac{1}{\Delta Smellbest}\times\frac{Maxgen}{i}\text{ (when }i\text{ is odd)} \tag{5}$$

$$r=\Delta Smellbest\times\frac{Maxgen-i}{Maxgen}\text{ (when }i\text{ is even)}, \tag{6}$$

where i is the current iteration number, $\Delta Smellbest$ is the amplitude for each optimization, and *Maxgen* is the maximum number of iterations.

3.3 Improved FOA–RBF neural network

Given that *spread* is the key factor of the RBF neural network, the improved FOA algorithm is adopted to optimize the value of *spread* in RBF. The critical point is to set *spread* value as the flavor concentration of the improved FOA algorithm and make the absolute value of the prediction square errors as the decisive function of the flavor concentration; thus, the process is as follows (Xu, 2015):

Step 1: Set the initial parameters of the fruit flies groups: Random initial position (X_0, Y_0), population size *sizepop* $= 40$, and the number of iterations *Maxgen* $= 20$.

Step 2: Give fruit flies the random direction γ (γ is the random number of [0, 1]) and the radius r (see Equations [4]–[6]) by their smell. After the ith iteration, the location of the quick flies is calculated as follows: (X_i, Yi). $(X_i = + r^*(\gamma-0.5), Y_i = Y_0 + r^*(\gamma-0.05))$.

Step 3: Estimate the distance Di between the location of the fruit flies and the original point $(0, 0)$ $\left(D_i = \sqrt{X_i^2 + Y_i^2} \right)$.

Step 4: Determine the value of the smell concentration S_i ($S_i = 1/D_i$, *spread* $= S_i$).

Step 5: Substitute the value into the function *newrbe*:

$net = newrbe(P,T,spread)$ % training of the RBF network;

$a = sim(net, P\_test)$ % prediction value of the RBF network;

$err = P\_test - a$ % the prediction error;

$Function_i = (err)^2$ % prediction square error.

Step 6: Compare all the decisive function values of the smell concentration corresponding to the locations of the ith fruit fly.

Step 7: Keep the best decisive function value and the position and carry on the iterative optimization until the cessation of iteration.

4   MIV-IMPROVED RBF NEURAL NETWORK MODEL FOR DAM DEFORMATION

4.1   Project summary

A water control project is located in the trunk stream of Kamchay River in Cambodia. The water control project consists of the high RCC gravity dam, five drainage sluice holes on the dam crest, diversion structures, power generation plant, and switching station. Besides, the width of crest is 6.0 m, the elevation of the crest is 153.00 m, and the one of the bottom is 41.00 m. Also, the maximum height of the dam is 112.00 m. In addition, by setting an extended wire, an inverted vertical wire, and the static level gages, the deformation monitoring statistical data are measured for analyzing the regularity of longitudinal, transverse, and vertical displacement in the different section and elevation of dam.

4.2   Deformation monitoring model

The deformation δ of the concrete gravity dam is under the action of water pressure, uplift pressure loads, temperature, etc. Based on these factors, deformation can be divided into hydraulic pressure component δ_H, temperature component δ_T, and aging component δ_θ (see Equation [7]) (Gu, 1990).

$$\delta = \delta_H + \delta_T + \delta_\theta \tag{7}$$

In Equation [7], hydraulic pressure component contains H, H^2, and H^3; temperature component contains T_1, T_2, T_3, and T_4; and aging component contains θ and $\ln(\theta)$.

As the dam is still in the initial impoundment period, the cracking and instability of the dam body are closely related to dam deformation. Thus, the deformation monitoring is an important part of the concrete gravity dam; meanwhile, the longitudinal displacement is the most important in the deformation monitoring (Wei, 2003). Thus, the longitudinal displacement and the related factors from January 10, 2012 to July 31, 2013 are selected as a sample data in the 1 # dam section of 153 m elevation to analyze the prediction results.

4.3   Screening predictor by the MIV algorithm

First, to avoid the effect of singular sample, all data of the sample are transformed to [0.1, 0.599] (see Equation [8]):

$$x^* = \frac{x - x_{min}}{x_{max} - x_{min}} \times 0.598 + 0.001 \qquad (8)$$

Second, the sample and the improved RBF are used to build the best RBF neural network (the best decisive function value of the smell concentration is 0.0118 after 20 iterations), where, now input neurons include H, H^2, H^3, T_1, T_2, T_3, T_4, θ, and $\ln(\theta)$, and output neurons include δ (longitudinal displacement). Lastly, utilizing the best RBF network, the MIV of every input neuron is calculated (see Table 1), and the flow diagram is shown in Figure 2. As shown in Table 1, H_3, T_4, H^2, and $\ln(\theta)$ have more effect on δ. Finally, these four factors are chosen as new input neurons in the RBF network.

4.4   Calculation results

After the screening of MIV algorithm, those four new input neurons and the displacement are reorganized as a new sample, which is divided into two parts including the training and the predicting. Whereas, the training part contains 271 groups and the predicting part contains 25 groups.

To verify the effectiveness of the MIV-improved RBF network model, the multiple linear regression, the conventional RBF network, and the MIV–RBF network models are also built. Besides, the multiple linear regression equation is given (see Equation [9], and the *spread* values in the conventional RBF network is set as 1 (the default). Prediction results of different models are shown in Figure 3, and the prediction errors of these models are presented in Table 2.

$$\delta = 8.1706 - 0.0089484 * H - 0.010854H^2 + 0.00066328H^3 + 0.031661T_1$$
$$-0.26573T_2 - 0.033723T3 - 0.023756T4 - 0.00022405\theta + 0.10464\ln(\theta), \qquad (9)$$

where H = water level; T_1, $T2$, $T3$, and $T4$ = temperature; and θ and $\ln(\theta)$ = aging.

Table 1.   Abstract MIV values of every factors.

Factors	H	H^2	H^3	θ	$\ln(\theta)$	T_1	T_2	T_3	T_4
Abstract MIV	0.0058	0.307	0.37	0.0045	0.142	0.0074	0.343	0.0078	0.0046
Order	7	3	1	9	4	6	2	5	8

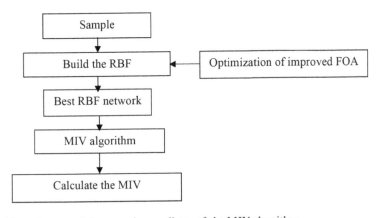

Figure 2.   Flow diagram of the screening predictor of the MIV algorithm.

Table 2. Prediction errors of different models.

Items	Multiple linear regression model	Conventional RBF neural network model	MIV–RBF neural network model	MIV-improved RBF neural network model
Average relative error	−50.25%	−33.72%	−9.22%	−4.14%
Mean square error	0.335	0.506	0.111	0.035

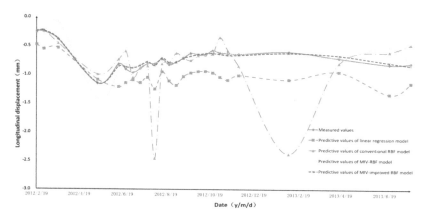

Figure 3. Prediction results of different models.

Figure 3 shows that the conventional RBF model deviates from the actual measured value; the multiple linear regression model meets the need of the real trend, but overall there is also a big deviation; the MIV–RBF and the MIV-improved RBF models are superior to other models, with a good trend and a stable regression; in addition, the fitting degree of the MIV-improved RBF model is high, and its prediction result is better.

At the same time, the average relative errors of the multiple linear regression, the conventional RBF network, the MIV–RBF network, and the MIV-improved RBF network models are 50.25, 33.72, 9.22, and 4.14%, respectively; the mean square errors of ones are 0.335, 0.506, 0.111, and 0.035, separately. Therefore, after being screened by the MIV and optimized by the improved FOA, the prediction accuracy of the conventional RBF network model increases obviously, and the prediction error is small, which can be used in the forecast analysis of dam deformation.

5 CONCLUSION

Based on the comparisons of different models, the following conclusions can be drawn:

1. The key point of dam deformation prediction is not only the screening of forecasting factors, but also the establishment of the optimal monitoring model.
2. The calculation results of the project example show that the MIV-improved RBF neural network model is equipped with the forecast stability and a high precision, and can be applied to dam deformation monitoring.

ACKNOWLEDGMENTS

This research was financially supported by the National Science Foundation of China (51279050) and the Ministry of Water Resources of China (201501033).

REFERENCES

Cao Xinrong, Lou Zhanghua,Sun Hongwei. Time series prediction model of vertical displacement of watergate based on RBF neural network [J]. Three Gorges University Journal: Natural Science Edition, 2010, 32(5): 17–19. (in Chinese).

Gu Chongshi, Wu Zhongru,Safety monitoring of dams and dam foundations-theories and method and their application[M]. Hohai University Press, 1990.

Huang Xiaofei, Gu Hao. Ant colony clustering radial basis function neural network model for safety monitoring of dam deformation [J].Three Gorges University Journal: Natural Science Edition, 2014, 36(6): 33–36. (in Chinese).

Li Dong, ZHANG Wenyu. Stock price prediction based on ELM and FOA [J]. Computer Engineering and Applications,2014,18:14–18+32. (in Chinese).

Lv Beibei, Yang Yuanfei. Application of PSO-RBF in dam deformation monitoring [J]. Water Resources and Power, 2012, 30(8): 77–79. (in Chinese).

Pan W T. A new fruit fly optimization algorithm: Taking the financial distress model as an example[J]. Knowledge-Based System,2012, 26(1): 69–74.

Wei Derong. On working out dam safety monitoring index[J]. Dam and Safety, 2003 (6): 24–28. (in Chinese) Zhang Xiaochun, Xu Hui, Deng Nianwu,et al. Application of a radial basis function neural network model to data processing technique of dam safety monitoring [J]. Journal of Wuhan University: Engineering Science, 2003, 36(2): 33–36. (in Chinese).

Xu Fuqiang, LIU Xiangguo. Variables screening methods based on the optimization of RBF neural network[J]. Computer Systems and Applications, 2012 (3): 206–208. (in Chinese).

Xu Guobin, Han Wenwen, Wang Haijun et al.Prediction of vibration response of powerhouse structures caused by flow discharge [J].Journal of Tianjin Normal University: Natural Science Edition, 2015, 48(3): 196–202. (in Chinese).

Hydraulic Engineering IV – Xie (Ed.)

The study of probability flood forecast based on AREM and Bayesian model

Zhiyuan Yin
Hubei Key Laboratory for Heavy Rain Monitoring and Warning Research, Institute of Heavy Rain, China Meteorological Administration, Wuhan, China

Fang Yang
Meteorological Information and Technology Support Center of Hubei Province, Wuhan, China

Tao Peng & Tieyuan Shen
Hubei Key Laboratory for Heavy Rain Monitoring and Warning Research, Institute of Heavy Rain, China Meteorological Administration, Wuhan, China

ABSTRACT: To improve the accuracy of precipitation in the forecast period, according to the hourly precipitation, the flow data of the flood season during the period of 2006–2008, and the corresponding 0–60 h hourly AREM forecasting precipitation provided by the Institute of Heavy Rain (IHR) in WuHan, in this paper, we took the Zhanghe reservoir basin in Hubei Province as the research object, introduced the Bayesian probabilistic model to correct precipitation forecasted by the AREM model, and analyzed the revised and unrevised precipitation. Then, both forecasting precipitations of the AREM model were inputted into the Xin'An-jiang hydrology flood forecast model. The results showed that the revised forecasting precipitation had the higher accuracy than the unrevised forecasting precipitation, and the root mean square error was reduced by not more than 10%. The average flood deterministic coefficient increases by 10.66%, and the average relative error of peak reduces by 3.05% in the verification period. It improved the precision of the flood forecast to some extent.

1 INTRODUCTION

Flood forecasting was mainly concerned with the flood peak and the total amount of water in rivers or basin reservoir, and the meteorological and the geographical factors were the mainly influences of the results. At present, the flood forecasting model accepted many kinds of information such as weather, hydrology, and so on. In this inputting information, precipitation was one of the most important information. In recent years, with the rapid development of numerical forecast technology, the accuracy of rainstorm forecasting has been improved; however, hydrology and the meteorology scholars are more and more concerned about the precipitation in the forecasting period. Because of the great uncertainty of precipitation during the forecast period, many experts and scholars have conducted a lot of research on how to accurately describe the precipitation in the forecasting period. S. X. Wang pointed out that the probabilistic hydrological forecast based on the Bayesian theory considering various aspects of uncertainty could meet the need of optimization decision and described the uncertainty of the hydrological forecasting using the distribution function. According to the analysis of the history of precipitation data on Sichuan Province, C. P. Chen, using the regional scale collection forecasting products to calculate the prior probability and establish the likelihood function, had done the Bayesian test of the torrential rain forecasting in Sichuan Province and achieved satisfactory results. However, although the precipitation products provided by the current numerical forecast could meet the requirement of the hydrological forecast, the forecasting accuracy was still to improve further. Therefore, this article introduced the Bayesian probability statistical model to revise the forecasting precipitation of the AREM mesoscale numerical model. In order to improve the accuracy of reservoir flood forecasting, we took the Zhanghe reservoir basin in Hubei Province as the research object. Then, we inputted the revised numerical model product into the hydrological forecasting model.

2 THE STUDY METHODS

Zhanghe reservoir is located in the tributary of Juzhang River, which is in the north shore of the Yangtze River; the reservoir control basin area is 2980 km²; the bearing rain area is 2212 km²; the total capacity is 20.35×10^8 m³; and the multi-year average precipitation was 1003.6 mm. As the Zhanghe reservoir is located in the middle and lower reaches of the Yangtze River, we have chosen the Xin'anJiang hydrological model that is suitable for flood forecasting of humid regions and the AREM mesoscale numerical model that has a strong ability to forecast rainfall in the middle and lower reaches of the Yangtze River as the basis for this research.

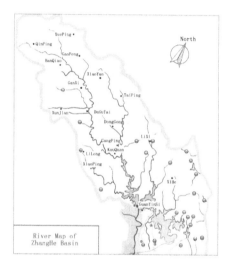

Figure 1.   Map of the research basin.

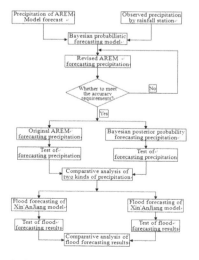

Figure 2.   Flow chart of the technical route.

46

When Krzysztofohicz proposed the Bayesian forecasting processor Bayesian Processor of Forecasts (BPF), he defined two variables. One was the forecasting result S and the other was the observation value of discrete time H. The values of these two variables were expressed by s and h, respectively. The probability prediction process is shown in Figure 3 (where k is the prediction time, and HK is the value that would be forecasted).

From Figure 3, we can see that the known variables are $h_1, h_2,..., h_{k-1}$ and $s_1, s_2,..., s_{k-1}, s_k$, which can be used to calculate the unknown variable h_k at the time of k. The specific calculation formula can be obtained as follows:

$$\Phi_k(h_k \mid h_1,...,h_{k-1},s_1,...,s_k) = \frac{f_k(s_1,...,s_k \mid h_1,...,h_k)g_k(h_1,...,h_k)}{\int f_k(s_1,...,s_k \mid h_1,...,h_k)g_k(h_1,...,h_k)dh_k}. \tag{1}$$

In general, the prior distribution and the likelihood function of BPF are usually calculated based on the linear normal assumption. However, this assumption had great limitations in practical applications. Then, Krzysztofohicz and Kelly proposed a method to deal with the uncertainty of hydrology, which was based on the Bayesian formula and the sub-Gaussian distribution function. The core content of the sub-Gaussian distribution function was Normal Quantile Transform (NQT). NQT was the marginal distribution function of known variables, which obeyed the normal strictly increasing distribution. Appointing Q as the standard normal distribution (the same as below), the converted normal quantile of h_k and s_k were as follows:

$$W_k = Q^{-1}(\Gamma(H_k)), k = 0,1,...,K \tag{2}$$

$$X_k = Q^{-1}(\bar{\Lambda}(S_k)), k = 1,...,K \tag{3}$$

In Formulas (2) and (3), W_k and X_k are normal quantile of H_k and S_k, respectively; Γ and $\bar{\Lambda}$ are, respectively, the marginal distribution function of H_k and S_k. After obtaining the normal quantile W_k and X_k, the prior distribution and the likelihood function could be constructed based on the Bayesian method in the transformed space. According to the data obtained, H_k is defined as the hourly average observed precipitation of the basin and S_k as the hourly average AREM forecasting precipitation of the basin.

1. Prior distribution: assuming that W_k obeyed the following linear relationship:

$$W_k = cW_{k-1} + \Xi, \tag{4}$$

where c is the coefficient and Ξ is the residual series, which does not depend on the parameter W_{k-1} and follow the normal distribution of $N(0,1-c^2)$.

2. Likelihood function: assuming that X_k, W_k, and H_0 obeyed the following linear relationship:

$$X_k = a_kW_k + d_kW_0 + b_k + \Theta_k, \tag{5}$$

where a_k, b_k, and d_k are the coefficients and Θ_k is the residual series, which does not depend on the parameters (W_k, W_0) and follow the normal distribution of $N(0,\delta_k^2)$.

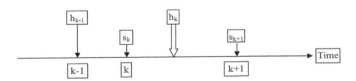

Figure 3.   Flow chart of the probabilistic forecasting.

After deriving a series of formulas and space conversion, the original space of the sub-Gaussian posterior distribution function was obtained. In this paper, it was based on the AREM forecasting probability precipitation:

$$\Phi_k(h_k \mid s_k, h_0) = Q\left[\frac{Q^{-1}(\Gamma(h_k)) - A_k Q^{-1}(\bar{\Lambda}_k(s_k)) - D_k Q^{-1}(\Gamma(h_0)) - B_k}{T_k}\right], \tag{6}$$

where $t_k^2 = 1 - c^{2k}$, $A_k = \dfrac{a_k t_k^2}{a_k t_k^2 + \delta_k^2}$, $B_k = \dfrac{-a_k b_k t_k^2}{a_k t_k^2 + \delta_k^2}$, $D_k = \dfrac{c^k \delta_k^2 - a_k d_k t_k^2}{a_k t_k^2 + \delta_k^2}$, and $T_k = \dfrac{t_k^2 \delta_k^2}{a_k t_k^2 + \delta_k^2}$.

3 THE ANALYSIS OF THE EXAMPLE

Several mathematical models such as gamma distribution, log-Pearson distribution, kappa distribution, log-logistic distribution, and log-Weibull distribution were selected for the marginal distribution function of H_k and S_k; however, through the preliminary analysis and comparison, the log-Weibull distribution was chosen as a suitable model in this paper. The expression of the distribution function is given as follows:

$$F(h_k) = 1 - \exp[-(\frac{h_k - t_0}{\eta - t_0})^m], \tag{7}$$

where t_0, η, and m are the coefficients and satisfied $h_k \geq t_0, \eta > 0, m > 0$.

According to the history data and the least squares method, the parameters can be calculated (see Table 1).The probability value of the empirical point was calculated by the formula $P = m/(N+1)$, from which the log-Weibull distribution can very well be described as the probability distribution of the measured precipitation and the forecasting precipitation. The correlation coefficients were, respectively, 0.9983 and 0.9965 (see Figure 4), which were almost complete with the actual distribution. Therefore, the log-Weibull distribution was reasonable for the marginal distribution function of Γ and $\bar{\Lambda}$.

Table 1. Parameters of the marginal distribution function.

Distribution function	t_0	η	m
Γ	0.13	2.6182	0.6944
$\bar{\Lambda}$	0.14	2.5064	0.5141

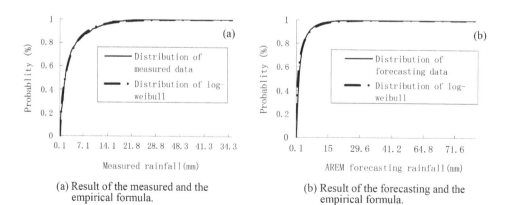

(a) Result of the measured and the empirical formula.

(b) Result of the forecasting and the empirical formula.

Figure 4. Comparison of the empirical point and the log-Weibull distribution.

48

According to the historical data to determine the sample {h0, h1} of two adjacent hours, through the normal quantile of the sub-Gaussian space to get the sample {w0, w1}, the results of the correlation analysis of two groups of samples were shown in Figure 5. After calculating, the parameter c was 0.8705, correlation coefficient of {h0, h1} was 0.6718, and the correlation coefficient of {w0, w1} was 0.8618. The historical data converted by the normal quantile transformation had obvious improvement in sample correlations. Therefore, the use of the linear hypothesis assumption in the prior distribution was more reasonable than the original data in the sub-Gaussian model.

Similar to the prior distribution, the sample $\{s_1, h_0, h_1\}$ converted by the normal quantile transformation became the sample $\{x_1, w_0, w_1\}$; the associated parameter values are listed in Table 2. From Figure 6, we can see that if the linear equations were directly established, the correlation coefficient was only 0.6791 according to the original sample, but after the quantile transformation, the sample correlation coefficient compared with the previous results significantly increased and reached to 0.8122 (see Figure 6). Therefore, it was more reasonable to establish the linear equation in the converted sample than in the original sample.

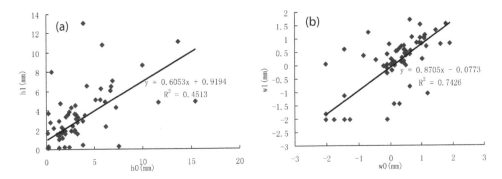

Figure 5. Correlation analysis of the prior distribution samples.

Table 2. Parameters of the likelihood function.

k	a_k	b_k	d_k	δ_k^2
1	0.4673	−0.0517	0.2879	0.0675

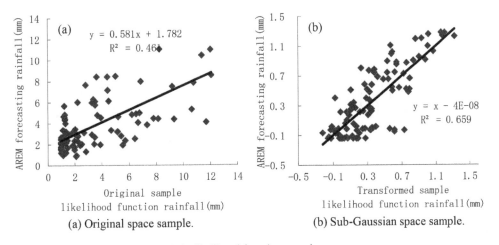

(a) Original space sample.   (b) Sub-Gaussian space sample.

Figure 6. Correlation analysis of the likelihood function samples.

49

The difference between the prior distribution and the model forecasting shown in Figure 7(a) is larger than that shown in Figure 7(b), but the changing trend of the result of Bayesian forecasting was always consistent with the result of the AREM forecasting model, and the magnitude of both precipitations was very close. According to calculation of the root mean square error, the Bayesian forecasting model had the minimum error, and the second was the AREM forecasting model. The prior distributions had the maximum error, but the extent of the root mean square error decreases, which always remains below 10%. According to the result of the AREM forecasting precipitation model, when the confidence interval [−0.95, 0.95] is given in the sub-Gaussian space and is converted by NQT, it could obtain the probabilistic precipitation forecasting interval [$p_{-0.95}$, $p_{0.95}$] in the original space. The intervals $p_{-0.95}$ and $p_{0.95}$, which were the precipitation sequence changing with time, were, respectively, expressed by {$P_{-0.95}$} and {$P_{0.95}$}. The corresponding flow result was expressed by {$Q_{-0.95}$} and {$Q_{0.95}$}. The result of the flood forecasting under the corresponding precipitation is shown in Figures (c) and (d). The measured flood curve was sandwiched between {$Q_{-0.95}$} and {$Q_{0.95}$}, and the effect of the mean probability forecasting (Q_0) was better than that of {$Q_{-0.95}$} and {$Q_{0.95}$}. In other words, under the condition of the given confidence interval, the probabilistic flood forecasting model could not only give the possible range [$Q_{-0.95}$, $Q_{0.95}$] of flood changing but also provide a credible forecasting reference result Q_0.

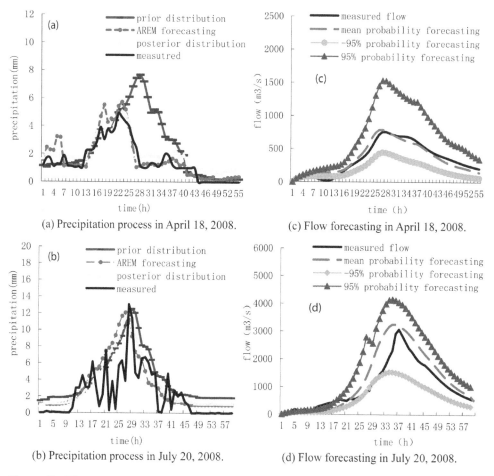

(a) Precipitation process in April 18, 2008.

(c) Flow forecasting in April 18, 2008.

(b) Precipitation process in July 20, 2008.

(d) Flow forecasting in July 20, 2008.

Figure 7. The Bayesian probabilistic precipitation forecasting and the corresponding flow forecasting.

4 CONCLUSION

1. When the sub-Gaussian model was used to convert the original samples, the linear assumption of the prior distribution and the likelihood function were more reasonable than ever.
2. The Bayesian forecasting model could improve the accuracy of the original forecasting result, but its accuracy depended on the selected numerical model and the improved effect was very limited (the root mean square error decreases by not more than 10%). If the precipitation forecasting accuracy of the AREM model is further enhanced or another more accurate numerical model is used, it will be helpful to improve the flood forecasting accuracy.
3. Using the Bayesian probabilistic model to revise the AREM forecasting precipitation and calculate the hydrological forecasting flow, the average deterministic coefficient increased by 10.66%, and the average peak relative error reduced by 3.05% compared with the unrevised AREM forecasting precipitation.

ACKNOWLEDGMENT

This work was supported by the Science and Technology Development Key Fund of Hubei Provincial Meteorological Bureau (2015Z02).

REFERENCES

Chen C.P. et al. Application of Sichuan Heavy Rainfall Ensemble Prediction Probability Products Based on Bayesian Method. Meteorological Monthly, 2010, 36(5):32–39.
Cui C.G. et al. The flood forecast test on QPF Coupling with hydrological model in flood season in mediu m and small catchment. Meteorological Monthly. 2010, 36(12):56–61.
Gong Y. Evaluation and analysis of the rainfall prediction of AREM in flood season of 2007.torrential rain and diasters. 2007, 26(4):372–380.
Kelly K.S., Krzysztofowicz R. Probability distributions for flood warning system [J].Water resour Res, 1994, 30(4):1145–1152.
Krzysztofowicz R. Bayesian model of forecasted time series [J].Water Resour Res, 1985, 21(5): 805–814.
Li J. et al. Precipitation forecast experiments of a mesoscale numerical model AREM. Meteorological Monthly. 2007.40(1):13–17.
Luo W.S. Anti-error researching of flood forecasting project of double mutual inflow reservoir. Engineering Journal of Wuhan University. 2006, 39(3):1–5.
Peng T. et al. Application of Radar QPE to Flood Forecast in Flood Season Hydrological Model. Meteorological Monthly. 2010, 36(12):50–55.
SL25-2000 hydrological forecast specification. Beijing: China Statistics Press, 2000:18–22.
Wang S.X. A brief introduction to Bayesian probabilistic hydrologic forecasting, Journal of China Hydrology, 2001, 21(5):33–34.
Wang Y.H. et al. A test for real-time precipitation forecast during 2002-flood season with AREM Model. Meteorological Monthly. 2005, 31(2):17–22.
Wu X.Z. Modern Bayesian Statistics. 2000, Beijing: China Statistics Press.
Zhang H.G. et al. Real-time flood updating model based on Bayesian method. Engineering journal of Wuhan university. 2005, 38(1):58–63.
Zhao R.J. A model of the Xin'An River and North Shaanxi watershed hydrological model simulation. Beijing: Water Conservancy and electric power press, 1984.

Hydraulic Engineering IV – Xie (Ed.)

Numerical simulation of knotless fishing nets in current

Hao Tang
College of Marine Sciences, Shanghai Ocean University, Shanghai, P.R. China

Liuxiong Xu
National Engineering Research Center for Oceanic Fisheries, Shanghai, P.R. China
Key Laboratory of Sustainable Exploitation of Oceanic Fisheries Resources, Shanghai, P.R. China
Shanghai Education Commission "Summit and Highland" Discipline Construction for Fisheries Sciences,
Shanghai, P.R. China

ABSTRACT: Netting is a basic element in many structures that is related to fishing nets, which can be described in a mathematical model as a flexible structure. Making fishing net visualization and improving numerical simulation accuracy are the main issues that need to be addressed in the field of gear design. In order to simulate the shape configuration of submerged plane nets in current, a mass-spring model is derived based on the lumped mass method. The explicit Euler method is used to solve simulation motion equations, and then the shape configuration of the net is obtained. The results of numerical simulation are presented to show the sinking behavior and equilibrium configuration of the knotless plane net in current.

1 INTRODUCTION

Netting is a basic element in many structures that is related to fishing nets, aquaculture nets, and other fish enclosures. The use of fishing gear requires better understanding of hydrodynamic characteristics of netting. Netting is different from other structures made up of rigid materials, which is generally considered as a flexible system. The flexible netting is subjected to various degrees of deformation during fishing operation. Therefore, it is difficult to accurately simulate the behavior of submerged flexible nets. In the past decades, two methods were used to analyze the performance of a flexible structure: flume tank experiment and computer numerical simulation. The numerical simulation is widely applied to design many fishing gears, including fish cages, trawls, purse seine, and other gears due to advantages such as reduced costs, time-saving, and visual design.

Previous studies have been carried out to construct physical models of fishing gear systems and to simulate underwater dynamic behavior of fishing nets. Aarsnes et al. (1990) proposed an analytical method to determine the shape and forces of fish cages. Théret (1993) used a three-dimensional model for designing fishing nets and developed simulation software to calculate the shapes and tension forces of trawls under constant currents. Hu et al. (1995) developed the trawl system model using the Lagrange equation. Wan et al. (2002) applied a nonlinear finite element method to simulate the configuration and force distribution of a plane net. Tsukrov et al. (2003) proposed a consistent finite element to model the hydrodynamic behavior of the net panel, and this new modeling technique has been applied to simulate the performance of fish cages in the actual ocean condition. Li et al. (2006) investigated the shape and tension distribution of submerged plane nets in current by using a 3D model, and confirmed that the net configuration predicted by the 3D model was in close agreement with the experimental results.

Making fishing net visualization and improving numerical simulation accuracy are the main issues that need to be addressed in the field of gear design. The lumped mass method has been proved to be effective in the fishing gear system. In this paper, the lumped mass is used to simulate the hydrodynamic and sinking behavior of plane nets in current. The purpose of our study is to visualize the shape configuration of a submerged plane net.

2 MATERIALS AND METHODS

2.1 Structural modeling

Netting can be described as a flexible structure in a mathematical model. The flexible structure is modeled on the mass-spring model, and the structure is described as a physical system divided into a finite number of mass points connected by springs (Figure 1). In order to build up the numerical model and simplify the mathematical model, the following assumptions are proposed: 1) there is only tension in the axis direction of mesh bar, and the tension is constant across the cross-section of the mesh bar; 2) the relative displacements of all points on the cross-section of the mesh bar are equal; 3) the cross-sectional area of the mesh bar remains constant during deformation; and 4) the netting twine is completely flexible and easily bent without resistance.

Therefore, it is assumed that the lumped points at each mesh bar have the fluid dynamic characteristics of cylindrical elements and that the coefficients of fluid force vary with the relative fluid velocity direction. We can estimate the net shape by calculating the displacements of these point masses under boundary and flow conditions.

2.2 Equation of motion

The basic equation describing the motion of each mass point is as follows:

$$\sum \vec{F} = (m + \Delta m)\vec{a} = \vec{F}_{int} + \vec{F}_{ext} \tag{1}$$

where $\sum \vec{F}$ is the composition vector of forces acting on the bar element, including hydrodynamic, gravity, buoyancy, and tension forces; m is the mass of mass point; Δm is the added mass; $\vec{a}$ is the acceleration vector; $\vec{F}_{int}$ is the internal force acting on the mass point; and $\vec{F}_{ext}$ is the external force applied to the mass point. The added mass of the mass points is given by the following equation:

$$\Delta m = \rho_w V_m C_m \tag{2}$$

where ρ_w is the density of seawater, V_m is the volume of the mass points, and C_m is the added mass coefficient.

The internal force is a function of fractional extension of the material extending along the line of the spring for each element of the bars. The internal forces can be represented by the following equation:

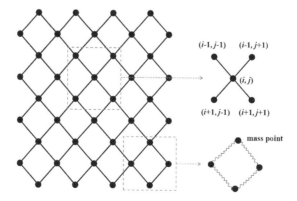

Figure 1. The schematic diagram of the mass-spring model.

$$F_{int} = \sum_{i=1}^{n} \frac{EA}{l_0}(l_1 - l_0)\bar{n}_l \tag{3}$$

where E is the elastic modulus of the bar, $A = \frac{\pi d^2}{4}$ is the effective area of the material, d is the diameter of the bar, l_0 is the initial length of the spring, l_1 is the deformed length, $\bar{n}_l$ is the unit vector along the line of spring, and n is the number of mass points.

The external forces $\bar{F}_{ext}$, including drag force $\bar{F}_D$, lift force $\bar{F}_L$, and gravity and buoyancy $\bar{F}_W$ (Figure 2), are as follows:

$$\bar{F}_{ext} = \bar{F}_D + \bar{F}_L + \bar{F}_W \tag{4}$$

The drag and lift forces are given by the following formulas:

$$\bar{F}_D = \frac{1}{2}C_D\rho_w SV^2\bar{n}_D \tag{5}$$

$$\bar{F}_L = \frac{1}{2}C_L\rho_w SV^2\bar{n}_L \tag{6}$$

where C_D is the drag coefficient, S is the projected area of mass point, V is the current velocity, C_L is the lift coefficient, $\bar{n}_D$ is the unit vector of the resultant velocity vector, and $\bar{n}_L$ is the unit vector of the lift force direction.

Buoyancy and gravity of mass points can be represented as follows:

$$\bar{F}_W = (\rho - \rho_w)\frac{\pi d^2}{4}l_0\bar{g} \tag{7}$$

where ρ is the density of the material and $\bar{g}$ is the gravitational acceleration.

2.3 Numerical method

A second-order nonlinear ordinary differential equation in the time domain is used in the numerical method. Thus, the acceleration of mass point i can be given by the following system of ordinary differential equations:

$$\begin{aligned} \ddot{x}_i &= f(x_i, y_i, z_i, \dot{x}_i, \dot{y}_i, \dot{z}_i, t) \\ \ddot{y}_i &= g(x_i, y_i, z_i, \dot{x}_i, \dot{y}_i, \dot{z}_i, t) \\ \ddot{z}_i &= h(x_i, y_i, z_i, \dot{x}_i, \dot{y}_i, \dot{z}_i, t) \end{aligned} \tag{8}$$

where x_i, y_i, and z_i are the location coordinates for the mass point i.

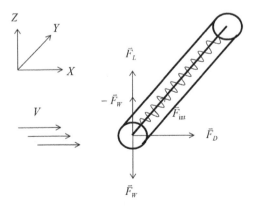

Figure 2. Definition of vectors for the spring system.

The net shape at each time step can be calculated numerically by solving an ordinary differential equation with a given initial value. The equations are solved by using the explicit Euler method. As it is a stiff equation, which is sensitive to algorithm, an implicit predictor–corrector method is used at the k-th iteration to ensure the precision and stability of numeric solution. The k-th iteration is acquired by Euler method as follows:

$$y_{k+1}^{(0)} = y_k + \Delta t \cdot f(t_k, y_k) \qquad (9)$$

The corrector procedure is again modified by using a trapezoid formula as follows:

$$
\begin{aligned}
y_{k+1}^{(1)} &= y_k + \frac{\Delta t}{2} \cdot \left[f(t_k, y_k) + f(t_{k+1}, y_{k+1}^{(0)}) \right] \\
y_{k+1}^{(i+1)} &= y_k + \frac{\Delta t}{2} \cdot \left[f(t_k, y_k) + f(t_{k+1}, y_{k+1}^{(i)}) \right]
\end{aligned}
\qquad (10)
$$

3   SIMULATION CASES

In order to obtain the shape configuration of the submerged plane net in current, a knotless plane net was used to perform the numerical simulation. Two cases are selected in this paper: 1) plane net of the top fixed and bottom weighted and 2) plane net with upper and bottom ends fixed. Figure 3 shows the sinking performance of a plane net in static water. The calculated configurations of the plane net in water, which are set vertically to a 0.2 m/s flow, are shown in Figures 4 and 5.

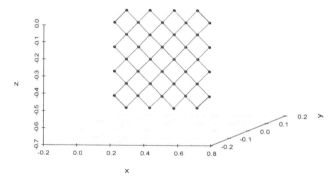

Figure 3.   The sinking behavior of a knotless net in static water.

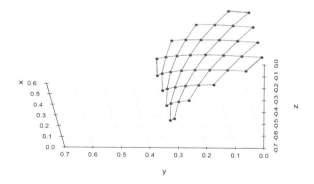

Figure 4.   The equilibrium configuration of the plane netting when the current speed is 0.2 m/s.

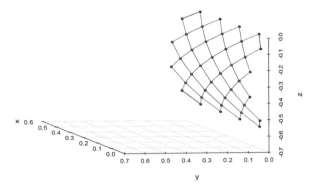

Figure 5. The equilibrium configuration of the plane netting in current when upper and bottom ends are fixed.

4 CONCLUSION

In order to simulate the shape configuration of a submerged plane net in current, in this paper, the mass-spring model was derived based on the lumped mass method. The results of numerical simulation were presented to show the sinking behavior and equilibrium configuration of a knotless plane net in current. In future, the lumped mass method can be used in a full 3D model of the fishing gear system.

ACKNOWLEDGMENTS

This study was financially supported by the National High Technology Research and Development Program of China (No. 2012 AA092302), the Shanghai Education Commission "Summit and Highland" Discipline Construction for Fisheries Sciences (No. B2-5005-13-0001-5), and the Open Funding for the Key Laboratory of Sustainable Exploitation of Oceanic Fisheries Resources (No. A-0209–15–0503–6).

REFERENCES

Aarsnes, J.V., Rudi, H., Løland, G., 1990. Current forces on cage, net deflection. In: Telford, Thomas (Ed.), Engineering in Offshore Fish Farming, Proceedings of the conference organized by the Institute of Civil Engineers, Glasgow, 17–18 October 1990, Thomas Telford Ltd. pp. 137–152.

Hu F., Matuda, K., Tokai, T., & Kanehiro, H. (1995). Dynamic Analysis of Midwater Trawl System by a Two-Dimensional Lumped Mass Method. Fisheries science, 61(2), 229–233.

Li, Y.C., Zhao, Y.P., Gui, F.K., & Teng, B. (2006). Numerical simulation of the hydrodynamic behaviour of submerged plane nets in current. Ocean Engineering, 33(17), 2352–2368.

Théret, F. (1993). Etude de l'équilibre de surfaces reticules places dans un courant uniforme (application aux chalets)(Ph. D. Thesis). Université de Nantes, Ecole Centrale de Nantes, France.

Tsukrov, I., Eroshkin, O., Fredriksson, D., Swift, M.R., Celikkol, B. 2003. Finite element modeling of net panels using a consistent net element. Ocean Eng, 30, 251–270.

Wan, R., Hu, F., & Tokai, T. (2002). A static analysis of the tension and configuration of submerged plane nets. Fisheries science, 68(4), 815–823.

Hydraulic Engineering IV – Xie (Ed.)
© 2016 Taylor & Francis Group, London, ISBN 978-1-138-02948-4

A multi-functional experiment setup for underwater soil cutting simulation and rake tooth optimization

Gongxun Liu, Guojun Hong & Qingbo Zhang
Key Laboratory of Dredging Technology of CCCC, CCCC National Engineer Research Center of Dredging Technology and Equipment Co., Ltd., Shanghai, China

Yehui Zhu & Liquan Xie
Department of Hydraulic Engineering, Tongji University, Shanghai, China

ABSTRACT: A multi-functional experiment setup was proposed in this paper for soil cutting simulation and rake tooth optimization. The setup includes the rigid soil bin, the rigid cover board, the spring loading system and the observation device of the pore water evaporation. The spring loading system, which included the rake tooth installation device, was fixed in the soil bin. The rake tooth was installed on the rake tooth installation device and the sloping angle of the rake tooth can be adjusted with the device. An observation device for pore water evaporation was set on the soil bin cover. The soil and saturation water was put into the bin before the experiment. The soil bin was sealed with the rigid cover board and was kept air tight during the experiment. The rake tooth was powered by the spring loading system in the cutting process. The pore water evaporation in the cutting process was observed through the observation device. The mechanism of cutting resistance in the dredging process could be revealed with series of experiments and the minimum traction force, the optimal dredging velocity and pore water evaporation can be tested quantitatively in this experiment setup.

1 INTRODUCTION

With the fast development of port and shipping industry, the need for expansion of harbors and waterways is rising rapidly, which stimulates the improvement on capacity and technology of dredgers. Drag suction dredgers are important parts of the dredging equipments, for their wide accommodation to soil types and high efficiency.

The cutting efficiency of rake teeth on a drag suction dredger is a control factor on the working efficiency of the dredger. Some researchers have made achievements on the cutting mechanism of rake tooth. Miedema (1982) mentioned that the cutting process could be divided into 5 stages, which involved the seepage in the soil, the negative pore pressure, the pore water evaporation due to excessive negative pore pressure and the dilatancy of soil in the cutting process. All these factors could influence the cutting efficiency and traction force of the rake tooth. Other achievements put emphasis on the traction force of the rake tooth in the cutting process (Reece, 1965; Hettiaratchi et al, 1974; Zhang et al, 1995; Hong, 2008), the deformation and failure patterns of soil (Godwin, 1977; He, 1998; Aluko, 2004), the pore pressure distribution in the soil (Miedema, 1984) and the numerical simulation of the cutting process (Zhou, 2000, Bagi, 2005, Zeng, 2006).

At present, the design of rake tooth still relies on the experience of technical workers. The research on the cutting mechanism of the rake tooth and the theoretical analysis on the cutting process are still limited, especially on the quantitative analysis on the pore water evaporation and the minimum traction force. Indoor model experiment is an ideal way for theoretical analysis on the cutting mechanism of the rake tooth and rake tooth designing. A multi-functional experiment setup was proposed in this paper for indoor soil cutting simulation and rake tooth optimization. The mechanism of cutting resistance in the dredging process could be revealed with series of experiments and the minimum traction force, the optimal dredging velocity and pore water evaporation can be tested quantitatively in this experiment setup.

2 SETUP STRUCTURE AND FUNCTION

The experiment setup was installed inside a rigid soil bin with a rigid cover board (see Fig. 1). It contained a spring loading system, which included rake tooth installation device on which the model rake tooth was fixed and an observation device for pore water evaporation which was installed on the cover board. The soil bin can be sealed with the rigid cover board and was kept air tight during the experiment.

The spring loading system included a pair of gliding bars, two gliding blocks, a pair of loading strings, fixing screws for the gliding blocks and a set of string fixing device. The gliding bars were parallel and were installed inside the soil bin. Two gliding blocks, a movable one and an immobile one, were linked with the loading string and are installed on the gliding bars. The fixing screws were installed on the gliding blocks; when they were tightened, the blocks were fixed on the gliding bars. The string fixing device (see Fig. 2) was installed on the side

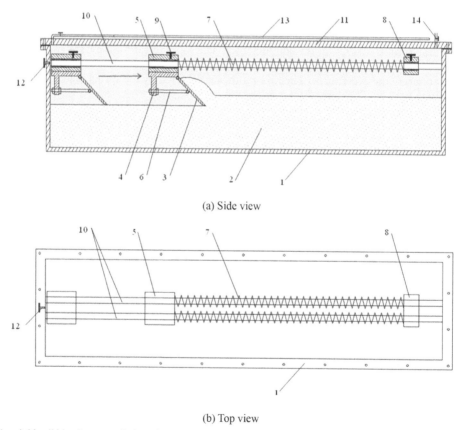

(a) Side view

(b) Top view

1 – rigid soil bin; 2 – test soil; 3 – rake tooth; 4 – rake tooth installation device; 5 – movable gliding block; 6 – rake tooth sloping angle adjustment screw; 7 – loading string; 8 – immobile gliding block; 9 – fixing screws for the gliding blocks; 10 – gliding bar; 11 – rigid cover board; 12 – string fixing device; 13 – pore water evaporation observation device; 14 – vacuum tube interface

Figure 1.   Layout of experiment setup.

of soil bin, which could fix the movable gliding block on its initial position. The string fixing device contained a steel wire, a fixing screw of the wire and a screw nut. The steel wire was linked to the movable gliding block on one end and could be trapped with the fixing screw on the other when the screw was tightened with the nut. The screw and the nut were installed on the outside wall of the soil bin. In the experiment, when the movable sliding block was set at the initial position on the left side of the soil bin, the fixing screws of the movable gliding block was tightened for temporary anchorage. Then the movable gliding block was fixed with the fixing device of the string, and thus the fixing screws could be removed and the string could be released from outside of the soil bin.

The rake tooth installation device (see Fig. 3) was installed on the movable gliding bar. The rake tooth was fixed on the device with two screws. One was on the top of the rake tooth, and the other was in the middle of the rake tooth. The sloping angle of the rake tooth could be adjusted with a screw bolt which was connected to the screw in the middle of the rake tooth. The sloping angle of the rake tooth could be set anywhere from 30 degrees to 120 degrees. The movement of the rake tooth was powered by the extended string. The cutting force and the cutting velocity could be adjusted freely by changing the initial distance between the two gliding blocks, i.e. the extension of the loading string. The maximum cutting velocity of the rake tooth was up to 1 m/s.

The observation device for pore water evaporation was a transparent thin tube which was fixed on the cover board of the soil bin. The tube went through the cover board and connected the inside of the soil bin. The tube was scaled on its side. The evaporation of pore water in the soil could be expressed quantitatively with the movement of water surface in the tube when the soil box was filled with water in the experiment.

The soil bin and the cover board could be linked with screw bolts. A sealing rubber ring was set between the cover board and the soil bin. The side walls and the cover board were made with toughened glass. A vacuum tube interface was set on the cover board for soil saturation.

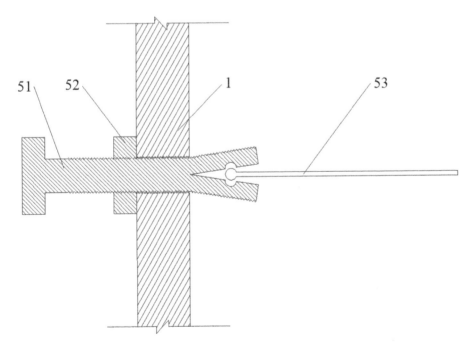

1 – rigid soil bin; 51 – fixing screw of wire; 52 – fixing screw nut; 53 – steel wire

Figure 2.   Details of the string fixing device.

3 EXPERIMENT STEPS AND POTENTIAL ACHIEVEMENTS

The basic steps for cutting process experiments of the rake tooth can be taken as follows:

Step 1: Put the test soil into the soil bin. The soil could be remolding soil or undisturbed soil taken from the in-situ tests.

Step 2: Set the sloping angle of the rake tooth with the adjustment screw bolt. Clear the soil above the designed soil surface. Clear the soil near the left side of the soil bin and above the bottom of rake tooth for the installation of the rake tooth and the movable gliding block.

Step 3: Install and anchor the movable gliding block on the left side of the soil bin with fixing screws. Install the loading string and immobile gliding block. Push the immobile gliding block to the right end of the gliding bar with a jack and fix it with the fixing screws.

Step 4: Install the string fixing device. Link the movable gliding block and the steel wire and fix the steel wire with the screw and the nut on the outside wall of the soil bin. Check the installation of the loading system and remove the fixing screws on the movable gliding block.

Step 5: Fill the soil bin with water and install the cover board. Vacuum the soil bin through the vacuum tube interface to saturate the test soil.

Step 6: Fill the soil bin with more water until the reading on the observation device for pore water evaporation is above the bottom line.

Step 7: Release the steel wire fixing screw on the outside wall of the soil bin, and the cutting process starts. Observe the cutting process, the reading on the observation device for pore water evaporation. In the cutting process, the rake tooth is powered by the restoring force of the string and the cutting velocity gradually slows down. When the rake tooth stops, the loading force of the string is the minimum traction force.

The following achievements may be made through this experiment.

1. Test of minimum traction force. The test would be a better guidance for the design of the rake tooth if the test soil was undisturbed.
2. Optimization of rake tooth cutting velocity. The object function of the optimization is the minimum energy cost when the cutting volume is the same.
3. Observation of pore water evaporation. The pore water in the soil evaporated when the negative pore pressure in the soil reaches vacuum point. The pore water in different types of soil reached evaporation in different conditions, which affects the characteristics of the cutting resistance.

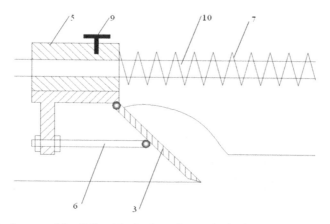

3 – rake tooth; 5 – movable gliding block; 6 – rake tooth sloping angle adjustment screw; 7 – loading string; 9 – fixing screws for the gliding blocks; 10 – gliding bar

Figure 3.   Details of the rake tooth installation device.

4. Cutting tests for critical cutting velocity of pore pressure evaporation. The critical velocity is reached at the initial point of pore pressure evaporation. The cutting resistance reaches a platform when the cutting velocity is over the critical value. The result of this critical value can be used for guidance in practical projects.
5. Researches on the mechanism of cutting resistance. Soil pressure gauges and pore pressure gauges can be utilized in the experiment to reveal the dynamic process of pressure development in the cutting process, which can be contributive to the analysis of cutting resistance mechanism.
6. New rake tooth research and development. Analysis and optimization on rake tooth efficiency can be conducted on the optimal cutting velocity, minimum traction force and critical cutting velocity based on different model rake teeth and certain sorts of soil.

4   CONCLUSION

A multi-functional experiment setup was proposed for soil cutting simulation and rake tooth optimization. This setup includes the rigid soil bin, the rigid cover board, the spring loading system and the observation device of the pore water evaporation. The soil bin can be sealed with the rigid cover board and was kept air tight during the experiment. The spring loading system, which included the rake tooth installation device, was fixed in the soil bin. In the cutting process, the rake tooth was powered by the spring loading system. The rake tooth was installed on the rake tooth installation device and the sloping angle of the rake tooth can be adjusted with the device. An observation device for pore water evaporation was set on the soil bin cover. The mechanism of cutting resistance in the dredging process could be revealed with series of experiments and the minimum traction force, the critical cutting velocity of pore pressure evaporation, the optimal dredging velocity and pore water evaporation can be tested quantitatively in this experiment setup.

REFERENCES

Aluko, Chandler H.W. Characterisation and Modelling of Brittle Fracture in Two-dimensional Soil Cutting, 2004, Vol.88(3): 369–381.
Bagi K. Methods to Generate Radom Dense Arragement of Particles. Paris: Institute Henri Poincare, Session on Granular Matter Thematic meeting on numerical simulations, 2005.
Godwin, R.J., Spoor G. Soil Failure with Narrow Tines J Agric Eng Res, 1977, 22(4): 213–228.
He J., Vlasblom W.J. Modelling of saturated sand cutting with large rake angle. 15th world dredging congress, June 1998, Las Vegas, Nevada, USA.
Hettiaratchi D.R.P, Reece A.R. 1974. The calculation of passive soil resistance. Geotechnique, 24(3): 45–67.
Hong G., Lin F., Wang J. et al. Experimental study on cutting of saturated hard compacted soil with draghead teeth. Waterway engineering, 2008, (8): 98–104.
Miedema S.A. De interactie tussen snijkop en grond in zeegang. Proc. Baggerdag 19/11/1982, T.H. Delft, 1982.
Miedema, S.A. Mathematische modelvorming t.a.v. een snijkopzuiger in zeegang. T.H. Delft 1984.
Reece A.R., 1965. The fundamental equation of earth-moving mechanics, Symp. on Earth-Moving Machinery, Proc. Inst. Mech. Eng: 16–22.
Zeng Y. Microscopic mechanics of soil failure and PFC numerical simulation. Shanghai: College of civil engineering, 2006.
Zhang J., Kushwaha R.L. A modified model to predict soil cutting resistance. Soil & Tillage Research (34): 1995, 157–168.
Zhou J., Chi Y., Chi, Y. et al. Simulation of biaxial test on sand by particle flow code. Chinese Journal of Geotechnical Engineering, 2000, 22(6): 701–704.

Hydraulic Engineering IV – Xie (Ed.)
© 2016 Taylor & Francis Group, London, ISBN 978-1-138-02948-4

Changes in Benthic diatom communities at different locations in Pearl River downstream rivers of China and the applicability of major diatom indices on the water quality evaluation

R. Jiang & F. Yang
Pearl River Hydraulic Research Institute, Guangzhou, China

X.T. Wang
Environmental Monitoring Center of Pearl River Basin, Guangzhou, China

Z.X. Li
Pearl River Hydraulic Research Institute, Guangzhou, China

ABSTRACT: The study of changes in benthic diatom communities at different locations in the downstream rivers of Pearl River was reported. Two major diatom indices, i.e. the specific Pollution Sensitivity Index (IPS) and the Biological Diatom Index (IBD), were applied to evaluate the water quality in three downstream rivers of Pearl River. The results showed that changes in benthic diatom communities at different locations in the downstream rivers of Pearl River were significantly related to human and city interference. There was a good agreement between IPS and IBD for the samples obtained from Liuxi and Sui rivers. Therefore, either of these diatom indices is suitable for use in assessing the water quality of these two rivers. IBD is also suitable for assessing the water quality for the more polluted river, i.e. Guangzhou river.

1 INTRODUCTION

Pearl River, the largest river in southern China, is the major source of drinking water for dwellers in cities, towns, and villages in its basin area including Hong Kong and Macao. The Pearl River delta is one of the most heavily populated areas in China with an average density of 2722 persons/km².

In China, biological monitoring of rivers has been based mainly on the *Escherichia coli* saprobity and chlorophyll *a* of the seston, but in some districts, periphyton was also used (Wu et al. 2002). Recently, there has been a need to add to the monitoring programs about other features of algal communities, especially those based on diatoms. The assessment of water quality conditions in freshwater habitats with benthic diatoms has a long history. According to the patterns of diatom taxa distributions, a variety of diatom indices have been developed, including the Biological Diatom Index (IBD) (Coste et al. 2009), the trophic diatom index (TDI) (Kelly 1998), the specific Pollution Sensitivity Index (IPS) (Garcia et al. 2008), the Descy index (DESCY) (Descy 1979), and the European Economic Community index (EEC) (Descy et al. 1991). These indices have been widely used as the bio-indicators in the assessment of pollution in rivers and lakes. Among these, IPS and IBD have been considered to be the most suitable indices for European-wide applications in evaluating water quality (Prygiel et al. 1993; Eloranta et al. 2002).

Although some Chinese researchers used diatoms to monitor river pollution in the 1980s (Wu et al. 2002), the information and background data for the local diatom taxa are very limited and not well documented. Recently, a preliminary investigation has been conducted in the Dongjian River basin, one of the main river systems and the upstream of the Pearl River basin located in southern China, in which several of the above-mentioned diatom indices have been used in the evaluation (Deng et al. 2012). The study found that the results from IPS and IBD were quite consistent in evaluating the water quality of the river.

In general, a given diatom index has a good applicability for particular rivers. Therefore, the applicability of IPS and IBD indices in the evaluation of the other rivers in Pearl River basin seems promising (Deng et al. 2012). The outcome will be helpful in better understanding of the relationship between the diatom indices and the chemical and physical parameters of waters so that meaningful comparisons can be made between different sites in the Pearl River basin.

In the present work, we reported the changes in benthic diatom communities at different locations in the downstream rivers of Pearl River and the use of IPS and IBD diatom indices in the evaluation of the water quality of three downstream rivers of Pearl River, namely the Liuxi, Sui, and Guangzhou rivers. The major goal was to find the relationship between the diatom indices and the chemical/physical parameters related to the water quality of these samples, from which the applicability of these indices to the evaluation of the water quality of rivers can be determined.

2 MATERIALS AND METHODS

2.1 Locations of the rivers

The studied rivers, such as Liuxi, Sui, and Guangzhou rivers, are in the area of Pearl River basin (east longitude 23°06′ 31.89″–23°53′ 30.80″ and north latitude 111°47′ 38.95″–113°59′ 18.70″) in Guangdong Province in south China, where they are influenced by the tidal reach activity and enjoy a typical humid subtropical climate. Liuxi River is a tributary of the Pearl River, northeast of Guangzhou; the Sui River is located in the middle reaches of the Pearl River; and the Guangzhou River is located in the southeast of Guangzhou city close to the point where the river enters the South China Sea. The economic activity around the area of Guangzhou River is dominated by the urban and industrial complexes, while agriculture and tourism are dominant around the areas of Liuxi and Sui rivers.

2.2 Sample collection and analyses

Diatom sampling, preparation, and analysis: Diatoms were sampled at each of the site for the period from June 2009 to July 2011 at the sites shown in Figure 1.

Sample collection, preparation, and identification were conducted according to standard methods in EN (EN 13946. 2003; EN 14407. 2005). The diatom communities were analyzed by examining the values ranging between 400 and 450 from each site. According to Krammer and Lange-Bertalot (2000), the diatoms were identified at the lowest taxonomical level. The community counts were entered into OMNIDIA (version 3.1), the diatom database, and the index calculation tool (Lecointe 1993) from which the diatom indices were calculated. In this work, only the IPS and IBD were examined.

Chemical analyses: When collecting diatom samples, the corresponding water samples at the same locations were also collected. The chemical parameters in the water samples (specifically, Chemical Oxygen Demand (COD), ammonia nitrogen (NH_3-N), nitrate nitrogen (NO_3-N), phosphate (PO_3), silicate (SiO_4), nitrite nitrogen (NO_2-N), pH, Biochemical Oxygen Demand (BOD_5), and the physical parameters, specifically conductivity (Cond.), and Dissolved Oxygen (DO)) were measured by standard methods (APHA 1995).

Cluster analyses: We used the software package PRIMER (Plymouth Routines in Multivariate Ecological Research) developed at Plymouth Marine Laboratory, UK. The hierarchical agglomerate cluster method (Clarke & Gorley 2001) was applied on the basis of abundance means per station to differentiate the phytoplankton communities that utilize the Bray–Curtis similarity index. Data were previously log(x+1) transformed to remove the bias of highly abundant taxa (Thatje et al. 2003).

3 RESULTS

3.1 *Characteristics of the diatom assemblages*

The compositions of the assemblages changed at different sampling sites. *Achnanthidium minutissimum, Achnanthes minutissima, Nitzschia clausii, Nitzschia palea,* and *Navicula schroeteri* var. *symmetrica* (77 species) were the dominant genera in the upstream of Sui River. In Liuxi River, there was an abundance of *Achnanthes conspicua, Achnanthes minutissima* var. *saprophila, Cocconeis placentula* var. *euglypta, Cyclotella stelligera, Eolimna minima, Gomphonema parvulum, Melosira varians, Nitzschia clausii, Navicula cryptotenella,* and *Planothidium frequentissimum* (82 species). *Aulacoseira ambigua, Aulacoseira lirata, Aulacoseira granulata, Cyclotella meneghiniana, Luticola mutica, Melosira varians,* and *Nitzschia clausii* (63 species) were the dominant genera in the downstream of Guangzhou River.

3.2 *The changes in the diatom indices over time at the sampling sites*

Figure 2 shows that the profiles of IBD and IPS in the sample sampling sites from the Liuxi and Sui rivers were in good agreement, although the values of IBD were slightly higher than those of IPS. This observation is consistent with the work reported previously (Resende et al. 2010). In contrast, there is a poor agreement between the profiles of IBD and IPS from sites of the Guangzhou River, with the values of IBD being generally lower than those of IPS between 2009 and 2011.

It can be seen that the values of these diatom indices range from 8 to 20 for Liuxi River, 6 to 17 for Sui River, and 4 to 14 for Guangzhou River during the 3-year trial. As the lower values of the index indicate poorer water quality (Resende et al. 2010), it was not surprising that the indices for the Guangzhou River were lower, because the Guangzhou River area was dominated by urban and industrial complexes.

3.3 *Physical and chemical parameters of the corresponding waters*

The Pearl River originates from a slightly mountainous region. West River, East River, and North River with their tributaries run through the entire territory of the rich alluvial delta. The upstream water, from sites S1 to S11, contained low concentrations of nutrients and was nearly saturated with DO. As it flowed downstream, from sites G1 to G5 and L1 to

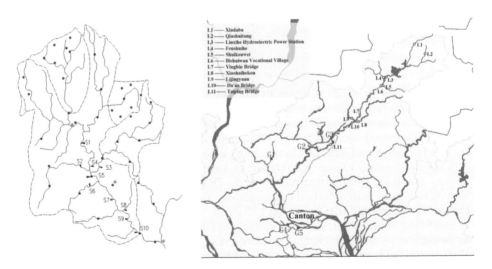

Figure 1. Sampling points for Sui (left), Liuxi, and Guangzhou rivers (right).

67

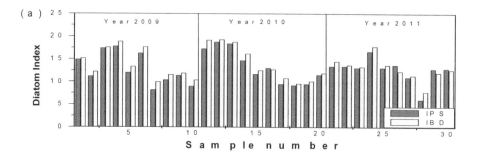

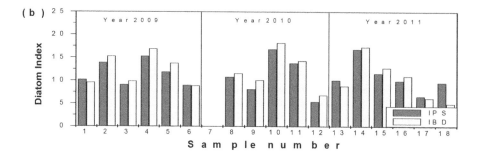

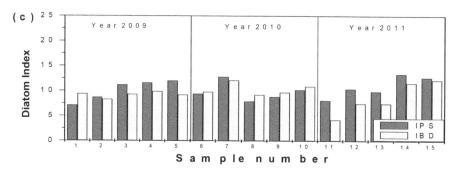

Figure 2. The profiles of IBD and IPS in the sampling sites at (a) Liuxi, (b) Sui, and (c) Guangzhou rivers.

L11, there was a progressive change in BOD_5, COD, turbidity, and ammonium. The environmental quality standards for surface water, i.e. GB3838–2002, were used in assessing the water quality of rivers. The data showed that the DO, BOD_5, and ammonium exceeded the standard V level in G1–G5 sites from 2009 to 2010, in which DO was 1.63 times the maximum standard value, and BOD_5 and ammonium were 2.17 and 1.65 times their corresponding maximum standard values. In 2011, after taking a series of actions for the protection of Guangzhou River from further pollution, the DO in G1–G5 sites increased to >2 mg/L (V level), and more than 60% BOD_5 samples met the standard requirement. Besides the exception for the DO and BOD_5 found at L10 during 2011, the water quality in all the other sites was found to be that of V level. The samples taken from the S1–S11 sites met the III level during the 3-year monitoring. Therefore, taking into consideration the water quality in these rivers, Sui River was the best, followed by Liuxi River, and Guangzhou River was the worst.

4 DISCUSSION

4.1 *Diatom distribution pattern changes related to water quality*

In Figure 3(a), the cluster analysis showed that the samples fell into three groups, and the similarity between two samples (i.e. S4 and S3) was 50%, while those among S1, S2, and S5–7 were ~38%. These samples show the same characteristic in diatom communities in space. The sample sections of S2, S3, S6, and S8 are located in the tributary of Sui River, near the river mouth. Moreover, the river section in S8 ran through a small city named Guangni, which is the dirtiest section in this river. It can be noted from Figure 3 that the diatom communities in S8 are only distantly related to the other communities.

In Figure 3(b), the cluster analysis shows that L7, L6, and L5 fall into a group with a similarity of 22%, while L7 and L8 have a very low similarity. L5 is located in the tributary of Liu River, L7 is before the city, and L8 is after the city. It can be seen from Figure 4 that the human activity in city area has a significant impact on the continuity of diatom communities.

In Figure 3(c), the cluster analysis showed that G1 and G2, G4 and G5 fell into two groups with a similarity of 41%. As the economic development in Guangzhou River area was more

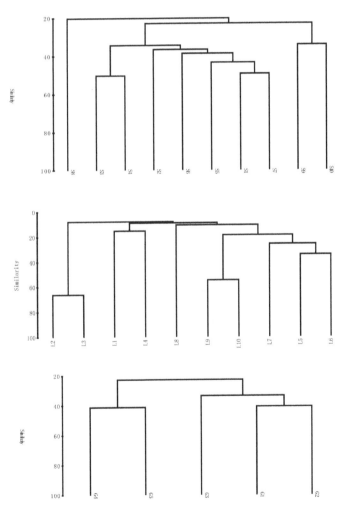

Figure 3. Dendrogram based on the cluster analysis of the abundance of diatom genera at the sampling sites in (a) Sui, (b) Liuxi, and (c) Guangzhou rivers.

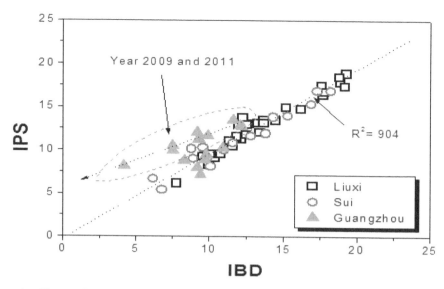

Figure 4. The correlation between IBD and IPS.

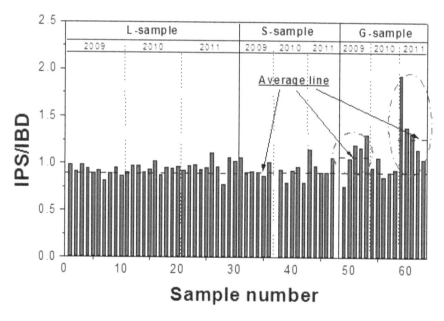

Figure 5. The correlation between IBD and IPS.

significant than that of the rural area, the sorts of the diatom species found in this river were less than those found in Sui and Liuxi rivers.

4.2 The relationship between the diatom index and major chemical parameters related to water quality

Figure 4 shows that the overall correlation between IBD and IPS indices obtained from these rivers over the 3-year study is reasonably good and fits the following equation:

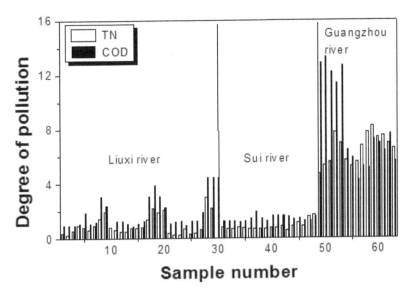

Figure 6.   The changes of TN and COD in the investigated rivers.

(1)

$$IPS = 0.955 \times IBD \,(\text{n} = 53, \text{R}^2 = 0.904)$$

The data that fell furthest from this line were those from the Guangzhou River in years 2009 and 2011. Although these data still seem to show a correlation between IPS and IBD, the relationship is different from that in Equation [1]. Figure 5 shows the ratios of IPS to IBD for all the samples in this study, highlighting by circling the samples that deviated from the linear relationship established in Figure 4 and Equation [1].

These results indicate that both IBD and IPS can be used as a good indicator of water quality in the Liuxi and Sui rivers. However, the use of these indices for evaluating water quality in the Guangzhou River is questionable.

Figure 6 shows the changes in two major indicators (i.e. TN and COD) of water quality in the three rivers from years 2009 to 2011. Clearly, the degree of water pollution in the Guangzhou River is the most serious among the rivers. Therefore, it is expected that the overall values of the diatom indices for Guangzhou River should be smaller than those of the Liuxi and Sui rivers. The data of the IBD shown in Figure 5 verify this expectation for the Guangzhou River, thereby demonstrating the suitability of IBD as a measure of water quality in the Guangzhou River.

5   CONCLUSION

The study showed that Sui River was the best, followed by Liuxi River, and Guangzhou River was the worst, with regard to the water quality in the period from 2009 to 2011. Diatom distribution pattern similarities in these three rivers were significantly related to human and city interference on the section of rivers. The results also showed that both IPS and IBD diatom indices were suitable to be used for the evaluation of water quality in the less polluted rivers, i.e. Liuxi and Sui rivers, in which there was a good agreement between IPS and IBD for the samples from these two rivers. However, there was a poor consistency between IPS and IBD for the Guangzhou River, a more polluted body of water. The analysis of the corresponding water quality data indicated that IBD should be more suitable for the quality assessment of Guangzhou River.

ACKNOWLEDGMENT

The authors acknowledge the financial support from the Natural Science Foundation of China (No. 51409287, 21576105), the Guangdong Province Hydraulic Science and Technology Innovation Project (No. 2012–15), and Ministry of Water Resources Public Welfare Industry Special Scientific Research Project (No. 201001021).

REFERENCES

APHA.1995. Standard Methods for the Examination of Water and Wastewater, 19th ed. American Public Health Association, Washington, DC.

Clarke, K.R. & Gorley, R.N. 2001. Primer v6: user manual/tutorial.PRIMER-E Ltd., Plymouth.

Coste, M., Boutry, S., Tison-Rosebery, J., & Delmas, F. 2009. Improvements of the Biological Diatom Index(BDI): Description and efficiency of the new version (BDI-2006). *Ecological Indications, 9,* 621–650.

Descy, J.P. 1979. A new approach to water quality estimation using diatoms. *Nova Hedwigia 64,* 305–323.

Descy, J.P., Coste, M. 1991. A test of methods for assessing water quality based on diatoms. *Verhandlungen Internationale Vereinigung für theoretische undangewandte Limnologie 24,* 2112–2116.

Deng, P.Y., Lei., Y.D., Liu, W. 2012. Exploration of benthic diatom indices to evaluate water quality in rivers in the Dongjiang basin, *Acta Ecologica Sinica 32,* 5014–5024.

Eloranta, P., & Soininen., J. 2002. Ecological status of some Finnish rivers evaluated using benthic diatom communities. *Journal of Applied Phycology 14,*1–7.

EN 13946. 2003. Water quality-Guidance standard for the routine sampling and pretreatment of benthic diatoms from rivers.

EN 14407. 2005. Water quality-Guidance standard for the identification, enumeration and interpretation of benthic diatom samples from running waters.

Garcia, L., Delgado, C., & Pardo, I. 2008. Seasonal changes of benthic communitier in a temporary stream of Ibiza (Balearic Islands). *Limnetica 27,* 259–272.

Kelly, M.G. 1998. Use of the trophic diatom index to monitor eutrophication in rivers. *Water Research 32,* 236–242.

Lecointe, A. 1993. software for taxonomy, calculation of diatom indices and inventories management, *Hydrobiologia 269/270,* 509–513.

Prygiel, J, & Coste, M. 1993. The assessment of water quality in the Artois-Picardie water basin (France) by the use of diatom indices. *Hydrobiologia 269 /270,* 343–349.

Resende, P.C., Resende., P. & Pardal., M. 2010. Use of biological indicators to assess water quality of the U1 River (Portugal). *Environmental Monitoring and Assessment 170,* 535–544.

Thatje, S., Schnack-Schiel, S. & Arntz, W.E. 2003. Developmental trade-offs in subanatarctic meroplankton communities and the enigma of low decapod diversity in high southern latitudes, Marine Ecology Progress Series 260:195–207

Wu, J.T., & Kow, & L.T. 2002. Applicability of a generic index for diatom assemblages to monitor pollution in the tropical Tsanwun, Taiwan. *Journal of Applied Phycology 14,* 63–69.

Hydraulic Engineering IV – Xie (Ed.)
© 2016 Taylor & Francis Group, London, ISBN 978-1-138-02948-4

Safety monitoring index of horizontal displacement based on abutment instability of the arch dam

C.J. Fu, X.W. Yao & H.Y. Shen
Large Dam Safety Supervision Center, National Energy Administration, Hangzhou, China

ABSTRACT: Safety performance of arch dams is currently conducted on the basis of monitoring data combined with engineering judgments. An advanced method for safety monitoring index was proposed based on the failure mode and overload capacity of an arch dam. By analyzing the potential failure mode of the arch dam, a three-dimensional finite element model was established to simulate the specific failure mode of abutment instability. The monitoring index was obtained during the gradual destruction process of the arch dam, which provides an important reference for the management of operation safety of the arch dam.

1 INTRODUCTION

Safety monitoring index is very important for the quick assessment of dam safety, which brings great convenience to the management of the dam. There are several methods to determine the monitoring index; the most commonly used ones are as follows (Wu et al. 1989). The confidence interval method is based on the mathematical model of the dam behavior and confidence belt related to probability. The small probability method establishes a set of monitoring data under the unfavorable combination of loads and obtains the safety monitoring index by determining the probability of failure. The above two methods are based on the mathematical statistics of monitoring data, without considering the structural characteristics of the dam. They are not applicable for the load combination, which does not actually occur. The limit state method is used to obtain the monitoring index under the condition that the structure resistance is equal to the force at the limit state. This method considers the structural characteristics of the dam but ignores the failure mode and the overload capacity.

This paper proposes an advanced method for safety monitoring index. This method is based on the analysis and simulation of the specific failure mode of an arch dam. Taking into consideration the overload capacity of the dam, safety monitoring index is determined by analyzing the behavior of the dam during gradual failure progress.

2 RESEARCH METHODOLOGIES

2.1 Analysis of failure modes of the arch dam

In order to analyze the failure mode of the arch dam, a description of the failure of a selection of arch dams is provided in this section. These case studies outline the causes that can lead to failure.

The failure of Malpasset dam (Veale & Davison, 2013) in France is probably the most investigated arch dam failure. It was a double-curvature arch dam of height 60 m and crest length 233 m. It was constructed in 1954 and failed in December 1959 due to the failure of left abutments. The left part of the dam slid along the shallow fracture zone under the arch thrust at a high water level. Ultimately, only the base of the dam remained on the bottom of the valley.

The Copper dam (Ru & Jiang, 1995) in the USA was an arch dam with 60 m height. High alkali cements were used during the construction. Alkali-aggregate reaction occurred in the operating period, leading to the irregular cracks on the upstream surface of the dam. Irreversible displacement of 140 mm was observed at the crest of the arch crown.

The Kölnbrein arch dam (Lombardi, 1991) in Austria was constructed in 1977. It had a maximum height of 200 m and a crest length of 626 m. The slenderness coefficient of this arch dam was relatively large because of the wide valley. Hence, the beam direction stress increased, causing a horizontal crack with the length of 100 m near the dam heel.

The Huaguangtan arch dam (Fu et al. 2014) in China was included as a recent example. The dam was 103.5 m high and had a crest length of about 227.9 m. In the second year after the impounding, the uplift and seepage of the monolith in the riverbed and left abutment increased quickly. It was found that there were cracks in the shallow part of the foundation. After analyzing, the cracks were formed from the inherent joint fissure in the rock mass of the dam foundations by tensile stresses that occurred during low-temperature and high water-level conditions.

The failure mechanism of the above-mentioned examples can be summarized as abutment instability, concrete deterioration, cracks, and dam heel failure. The statistical analysis of concrete arch dam incidents that occurred across countries revealed that abutment instability is the most important failure mechanism for the breakage of the arch dam. Therefore, the aim of this paper was to analyze safety monitoring index based on the failure mechanism of abutment instability.

2.2 *Monitoring index of the arch dam*

According to the statistics, damage to arch dams is a process that extends gradually from local damage to overall instability. This process is usually qualitatively defined at three stages: normal state, abnormal state, and failure state. The state characteristic value, a commonly used monitoring index, can be defined as the boundary value between adjacent states. Therefore, the safety monitoring index of the dam can be divided into two levels. The first level is the primary monitoring index, which is defined as the boundary value between the normal state and abnormal state of the dam, also called the warning value. The second level is the secondary monitoring index, which is defined as the boundary value between the abnormal state and failure state, also called the limited monitoring index (Wei, 2003).

The horizontal displacement of the dam is an important monitoring object. It is a comprehensive index that can be used to examine whether the dam is cracking, unstable, or over-stressing. Most of the arch dams are installed with plumbline groups, considered as the most reliable monitoring facilities, to monitor the horizontal displacement. Therefore, the horizontal displacement can be used as a direct and reliable monitoring object for the arch dam.

2.3 *Method for simulation of arch dam failure*

There are two kinds of simulation methods that can be used to analyze the failure of the arch dam: strength reserve method and over-loading method (Zhu, 2007). Based on the arch dam incidents, the main cause of failure is that the actual strength is much lower than the expected value. Moreover, the difference between the actual loads and design loads is not that obvious as strength reduction. Hence, it is realistic to simulate the failure of the arch dam by the method of strength reserve. The principle of the strength reduction method is illustrated as follows. The main material parameters (e.g. fault cohesive strength and friction coefficient) affecting abutment stability are multiplied by the strength reserve coefficient k_i. The initial value of the strength reserve coefficient k_0 is equal to 1.0 and reduced gradually to k_1, k_2,..., k_n. All k_i are applied to the model, respectively, until the failure of the arch dam occurs. Failure criteria of the arch dam obtained are as follows: 1) the displacement rate is increasing; 2) the displacement mutation is observed; and 3) the calculations do not converge.

2.4 The back analysis of material parameters

The actual elastic modulus that can increase the accuracy of the model can be obtained as follows (Wu, 2003): the actual elastic modulus of the dam and foundations are E_c and E_r, respectively, and the ratio of them is defined as $R = E_r/E_c$. When R is kept constant, the dam deformation caused by the water pressure varies inversely with the elastic modulus of the dam concrete E_c. The foundation deformation caused by water pressure also varies inversely with E_r. Let us assume that the values of elastic modulus are Ec' and E_r', and use them as input parameters for the structural analysis method to obtain the displacement component induced by water pressure, δ_H'. We then isolate the actual displacement induced by the water pressure from the measured data, δ_H. The actual elastic modulus of the dam and foundation rock can be described as follows:

$$E_c = E_c' \delta_H' / \delta_H \tag{1}$$

$$E_r = E_r' \delta_H' / \delta_H \tag{2}$$

2.5 Method for monitoring index based on abutment instability

The potential failure mode of abutment instability of the arch dam was obtained by analyzing the characteristics of the arch dam and the geological condition of abutment. A three-dimensional finite element model was developed in order to simulate the most likely failure mode of abutment instability. To increase the accuracy of the model, the back analysis of the main material parameters (e.g. the elastic modulus of the dam and foundation and the thermal expansion coefficient of dam concrete) was obtained based on field monitoring data. The safety states of the dam and the corresponding monitoring index were confirmed by failure criteria combined with the relationship between the development of dam behavior and the strength reserve coefficient.

3 CASE STUDY

3.1 General description of the dam

An arch dam, located in southwestern China and the subject of this case study, is a double-curvature thin concrete arch dam acting as a combined flood control measure and hydro-electric power facility. Its maximum height is 294.5 m and its width is 5.8 m at the crest and 17.49 m at the base of the central cantilever, respectively. The thickness-to-height ratio is 0.248. The arch length of the arch dam is 892.786 m, and the arc–height ratio is 3.035. The foundations of the dam are mainly composed of granite gneiss on the riverbed. Plumbline groups were installed on monoliths nos 4, 9, 15, 19, 22, 25, 29, 35, and 41 to monitor horizontal displacement. The construction of the dam began in January 2002 and the first impounding commenced in December 2005.

3.2 Analysis of failure modes of abutment

According to the design data, the E4 block with the minimum stability safety factor is considered as the most dangerous slip block. It ranges between the elevation of 1,050 and 1,150 m² in the right abutment. It is composed of three surfaces, which are as follows. The side surface is formed by the alteration zone (P1). The bottom surface cuts through flat joint sets (P2). The drawing open surface is formed by flat joint sets (P3) along the sideline of the upstream of dam foundation.

3.3 A three-dimensional finite element model

The finite element model was established by ABAQUS. Eight-node isoparametric elements were adopted in the model, which contained 4488 elements and 5810 nodes in the dam, and

7300 elements and 9564 nodes in the foundations. Weak structural planes (P1, P2, and P3) were described by Mohr–Coulomb strength theory (Barton & Choubey, 1977). The material parameters were generalized according to the design values, as given in Table 1.

Based on the back analysis method presented earlier, the site measurement data were used to obtain the actual material parameters of the arch dam, including the horizontal displacement measurement of 27 different parts of the dam and the horizontal displacement measurement of 18 different parts of the foundation. The results show that the inverse elastic modulus of the dam concrete (32 GPa) was larger than the design value (30 GPa). The inverse elastic modulus of the foundation was generalized to 19 GPa (design value 14.5–20 GPa). The inverse concrete temperature expansion ($9.1 \times 10^{-6}/°C$) was smaller than the design value ($8.20 \times 10^{-6}/°C$).

3.4 Numerical simulation of abutment instability

The finite element system was analyzed with inverse material parameters. The strength reserve coefficients of E4 block's weak structural planes (P1, P2, P3) were reduced from 1 to 0.001 stepwise. The combination of check flood level and high temperature led to the large abutment thrust, which was considered as the most unfavorable condition for the abutment stability of arch dams. Therefore, the scenario of check flood level and high temperature was applied for the model.

By observing the relationship between the development of dam behavior and the strength reserve coefficient, it is concluded that the monoliths of the dam near the right abutment is the most sensitive part to the slide of the E4 block. Considering the position of the displacement measure point, the sensitive spots of displacement were defined as elevation 1245 m in monolith no. 4, elevation 1190 m in monolith no. 4, and elevation 1245 m (dam crest) in monolith no. 9, which are marked as a, b, and c in Figure 1, respectively. The strength reserve coefficient corresponds to the deformation value of the sensitive spots listed in Table 2. Figure 2 shows the deformation development process with the decrease in the strength reserve coefficient. Figure 3 shows the key step of the deformation diagram of the finite element model with the deformation scale factor of 1000.

Table 1. Design value of the material parameter for back analysis.

Material	Elastic modulus GPa	Poisson's ratio	Unit weight t·m⁻³	Linear expansion coefficient /°C	Shear strength f	c'/MPa
Dam	30	0.18	2.4	8.20×10^{-6}	/	/
Foundation	14.5–20.0	0.28	2.7	/	/	/
P₁	3.5	0.35	/	/	0.68	0.24
P₂	6.0	0.30	/	/	0.92	0.68
P₃	6.0	0.30	/	/	0.92	0.68

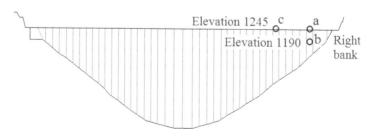

Figure 1. Position of the sensitive spots of displacement.

Table 2. Strength reserve coefficient corresponds to the deformation of sensitive spots.

Step	k_i	a mm	b mm	c mm	Step	k_i	a mm	b mm	c mm
1	1	17.35	14.78	52.46	9	0.25	20.00	17.59	57.31
2	0.9	17.42	14.85	52.60	**10**	**0.2**	**21.51**	**19.15**	**59.39**
3	0.8	17.50	14.94	52.77	11	0.1	45.68	42.55	84.49
4	0.7	17.53	14.96	52.80	12	0.05	74.91	70.80	114.40
5	0.6	17.75	15.20	53.30	13	0.03	91.82	87.18	132.00
6	0.5	17.97	15.44	53.78	**14**	**0.01**	**116.40**	**111.00**	**158.40**
7	0.4	18.40	15.88	54.58	15	0.005	122.70	116.70	166.20
8	0.3	19.24	16.79	56.11	16	0.001	139.50	131.40	187.80

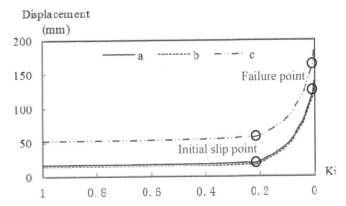

Figure 2. Deformation development process with the decrease in the strength reserve coefficient.

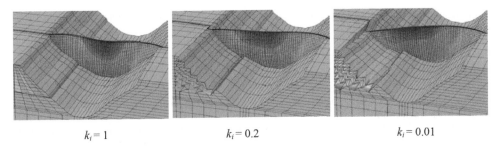

$k_i = 1$ $k_i = 0.2$ $k_i = 0.01$

Figure 3. Key step of the deformation diagram of the finite element model.

The result shows that the rate of displacement toward downstream is almost constant until the strength reserve coefficient reaches 0.2. It can be concluded that the safety factor for abutment stability is 5. Hence, the strength reserve coefficient of 0.2 is defined as the initial slip point. When the strength reserve coefficient reaches 0.01, the displacement mutation is observed, which means that the arch dam fails due to abutment instability. The strength reserve coefficient of 0.01 is defined as the failure point.

3.5 Safety monitoring index

Based on the basic principle of monitoring index presented earlier, the initial slip point is considered as the boundary value between the normal state and the abnormal state of the

Table 3. Safety monitoring index of deformation.

Sensitive spots	Corresponding measuring point	Primary monitoring index mm	Secondary monitoring index mm
A	A04-PL-01	21.51	116.40
B	A04-PL-02	19.15	111.00
C	A09-PL-01	59.39	158.40

dam, called the primary monitoring index. The failure point is defined as the boundary value between the abnormal state and failure state, called the secondary monitoring index. The safety monitoring indices of this arch dam are listed in Table 3.

4 CONCLUSION

The methodology described in this paper provides an effective means of obtaining the monitoring index of horizontal displacement based on abutment instability of the arch dam. The finite element model of an arch dam in China is established to simulate the failure progress of the dam by analyzing the potential failure mode. The safety states of the dam and the corresponding monitoring index are confirmed by failure criteria combined with the relationship between the development of dam behavior and the strength reserve coefficient. The key point of this method is the accurate analysis of failure mode and overload capacity of the arch dam. This results in a more reasonable criterion of the critical condition and a more practical safety monitoring index.

ACKNOWLEDGMENT

This work was supported by the Postdoctoral Research Funding Project of Zhejiang Province, China (BSH1502042).

REFERENCES

Barton, N., Choubey, V. 1977. *The shear strength of rock joints in theory and practice*. Rock Mechanics and Rock Engineering.
Fu, C.J., Yao, X.W., Li,T., et al. 2014. Investigation and evaluation of increasing uplift pressure in an arch dam: A case study of the Huaguangtan Dam. *KSCE Journal of Civil Engineering* 18(6): 1858–1867.
Lombardi, G. 1991. Koelnbrein dam: An unusual solution for an unusual problem. *International Water Power and Dam Construction* 43(6): 31–34.
Londe, P. 1987. The Malpasset Dam failure. *Engineering Geology* 24: 295–299.
Ru, N.H., Jiang, Z.S. 1995. *Arch dam: Accident and Safety of Large Dams*, Beijing: China waterpower press.
Veale, B., Davison, I. 2013. Estimation of gravity dam breach geometry. *NZSOLD/ANCOLD Conference*, 13–15 November.
Wei, D.R. 2003. On working out dam safety monitoring index, *Dam and Safety* (6):24–28.
Wu, Z.R., Lu, Y.Q. 1989. Feedback analysis of safety-monitoring values of dams with available in-situ measurements, *Journal of HoHai University*17 (6):29–36.
Wu, Z.R. 2003. The theory and application of safety monitoring of hydraulic structures, Nanjing: Hohai University press.
Zhu, B.F. 2007. Finite element whole course simulation and sequential strength reduction method for safety appraisal of concrete dams, *Journal of water resources and Hydropower Technology*38 (1):1–6.

Hydraulic Engineering IV – Xie (Ed.)
© 2016 Taylor & Francis Group, London, ISBN 978-1-138-02948-4

Risk prioritization of hydropower dams using risk informed decision making method

X.W. Yao, X.L. Zhang & C.J. Fu
Large Dam Safety Supervision Center, National Energy Administration, Hangzhou, China

ABSTRACT: Consideration of multiple risk factors is required to effectively evaluate and rank the relative risk performance of hydropower dams. This paper formulates comparison process of hydropower dams as a multi-criteria decision making model, and presents an approach by combining TOPSIS and grey correlation degree for solving the problem. This method can reflect both the relative closeness and geometrical similarity between the alternatives and the ideal solution. The methodology is applied to real cases in China to illustrate how the approach is used for the risk ranking problem of hydropower dams, which will assist dam owner and regulator in managing the hydropower dams in a simple and transparent way.

1 INTRODUCTION

At the end of 2014, there are 377 hydropower dams registered in the Large Dam Safety Supervision Center in China. The total installed capacity and total storage capacity of these registered hydropower dams are 139297.4MW and 3706.6×10^8 m³, accounted for 49.7% and 44.67% of that of all the reservoirs in China respectively. These dams play a very important role in people's daily life and the development of the whole country. With the consideration of the dynamic changes in the internal and external environments of the hydropower dams, the coexistence between the standards (codes) of safety and risk analysis techniques is not only beneficial but also necessary to ensure an effective aid to decision-making of dam safety management.

The risk evaluation of hydropower dams during its operating period is a complex process and is usually reflected by various risk factors. BC hydro (Hartford, 2014) developed a portfolio risk management framework of which the results are expressed in terms of vulnerability index scaling to a measure of the consequences of dam failure. Andersen et al. (2000) presents a risk index tool for the use of prioritizing the embankment dams based upon identifying deficiencies in the physical condition of the dam and rating their overall importance to the safety of the structure in terms of a postulated failure. The risk factors in these different risk prioritization methods reflect not only the possibility of dam failure but also the importance of dam break consequences. As a result, these risk factors and risk ranking provide useful information to the stakeholders of the dam, and reflect the dam's performance from various perspectives.

Multi-criteria Analysis (MA) is widely used in ranking or selecting one or more alternatives from a set of available alternatives with respect to multiple, usually conflicting criteria (Hwang & Yoon, 1981). MA provides an effective framework for hydropower dam comparison involving the evaluation of multiple risk factors. It can rank dams compared in terms of their overall risk performance. The aim of this paper is to develop the multi-criteria decision making method for ranking the registered hydropower dams in a simple and effective approach.

2 RESEARCH METHODOLOGY

2.1 The TOPSIS approach

In this paper, the concept of the approach used for solving the problem is based on the Technique for Order Preference by Similarity to Ideal Solution (TOPSIS). TOPSIS was initially presented by Lai et al. (1994), and Yoon and Hwang (1995). The concept of TOPSIS is that the most preferred alternative should not only have the shortest distance from the positive ideal solution, but also have the longest distance from the negative ideal solution.

In an risk prioritization problem, a set of hydropower dams (the alternatives $A = \{A_i, i = 1, 2, \cdots m\}$) is to be compared with respect to a set of risk factors (the criteria $C = \{C_j, i = 1, 2, \cdots n\}$). The performance rating of each dams A_i for each criterion C_j is a crisp value, and can be calculated from the available risk factor data. Therefore, an $m \times n$ performance matrix (the decision matrix X) for the problem can be obtained as

$$X = \begin{bmatrix} x_{11} & x_{12} & \cdots & x_{1n} \\ x_{21} & x_{22} & \cdots & x_{2n} \\ \cdots & \cdots & \cdots & \cdots \\ x_{m1} & x_{m2} & \cdots & x_{mn} \end{bmatrix} \tag{1}$$

where x_{ij} is a crisp value indicating the performance rating of each alternative A_i with regard to each criterion C_j. The decision matrix in Eq. (1) needs to be normalized for each criterion C_j ($j = 1, 2, \ldots, n$). The normalized value y_{ij} can be calculated as:

$$y_{ij} = \frac{x_{ij} - \min_j x_{ij}}{\max_j x_{ij} - \min_j x_{ij}}, y_{ij} \in Y^+ \tag{2}$$

$$y_{ij} = \frac{\max_j x_{ij} - x_{ij}}{\max_j x_{ij} - \min_j x_{ij}}, y_{ij} \in Y^- \tag{3}$$

where Y^+ is the set of criteria which have a positive impact in risk prioritization decision-making problems and Y^- is the set of criteria which have a negative impact.

The criteria weights are assigned by the decision makers and dam experts based on their own experiences, knowledge and perception of the problem via a preference elicitation technique such as the Analytic Hierarchy Process (AHP) (Saaty, 1995). As a consequence, the decision matrix is weighted by multiplying each column of the matrix by its associated criteria weight $W = (\omega_1, \omega_2, \cdots \omega_j)$ where ω_j is the weight of the jth criteria and $\sum_{j=1}^{n} \omega_j = 1$, $j = 1, 2, \cdots, n$.

The weighted normalized decision matrix representing the relative performance of the alternatives is obtained as

$$P = \begin{bmatrix} p_{11} & p_{12} & \cdots & p_{1n} \\ p_{21} & p_{22} & \cdots & p_{2n} \\ \cdots & \cdots & \cdots & \cdots \\ p_{m1} & p_{m2} & \cdots & p_{mn} \end{bmatrix} \tag{4}$$

where $p_{ij} = \omega_j y_{ij}$, $i = 1, 2, \cdots, m$, $j = 1, 2, \cdots, n$.

After determining the weighted normalized decision matrix of the alternatives, the next step is to aggregate them to produce an overall risk performance index for each alternative.

This aggregation process is on the basis of the positive ideal solution (A^+) and the negative ideal solution (A^-), which are defined, respectively, by

$$A^+ = \left(\max_i (p_{i1}), \max_i (p_{i2}), \cdots, \max_i (p_{in}) \right) = \left(p_1^+, p_2^+, \cdots, p_n^+ \right) \tag{5}$$

$$A^- = \left(\min_i (p_{i1}), \min_i (p_{i2}), \cdots, \min_i (p_{in}) \right) = \left(p_1^-, p_2^-, \cdots, p_n^- \right) \tag{6}$$

From Eqs. (4), (5) and (6), the Euclidean distances, between A_i and A^+, and between A_i and A^-, are calculated, respectively, as

$$D_i^+ = \left[\sum_{j=1}^{n} \left(p_j^+ - p_{ij} \right)^2 \right]^{1/2}, i = 1, 2, \cdots, m \tag{7}$$

$$D_i^- = \left[\sum_{j=1}^{n} \left(p_{ij} - p_j^- \right)^2 \right]^{1/2}, i = 1, 2, \cdots, m \tag{8}$$

An overall correlation degree for each alternative A_i ($i = 1, 2, \ldots, m$) is thus computed by the relative closeness to the ideal solution in TOPSIS. The relative closeness degree of the alternative A_i with respect to A^+ is defined as

$$D_i^* = \frac{D_i^-}{D_i^+ + D_i^-}, i = 1, 2, \cdots, m \tag{9}$$

The alternatives are ranked based on this relative correlation degree in a decreasing order. The larger the degree is, the higher the risk priority of the hydropower dams.

2.2 Decision making method based on TOPSIS and grey correlation degree

The traditional TOPSIS is on the basis of analyzing distance between alternatives and the ideal solution in solving multi-criteria decision problem. However TOPSIS can not reflect the situation variation of these alternatives. This paper combines TOPSIS and gray correlation degree to establish a new comprehensive correlation degree for the ranking of risk priority of alternatives. The main concept of grey correlation degree method is analyzing and comparing the similarity relationship between the geometrical curves of alternatives and of the ideal solution. This method chooses an alternative with the maximum grey correlation degree to the positive-ideal solution and the minimum grey correlation degree to the negative-ideal solution.

The gray correlation degree is determined based on the gray correlation coefficient which is calculated by

$$r_{ij}^+ = \frac{\min_i \min_j \Delta_i^+ (j) + \zeta \max_i \max_j \Delta_i^+ (j)}{\Delta_i^+ (j) + \zeta \max_i \max_j \Delta_i^+ (j)} \tag{10}$$

$$r_{ij}^- = \frac{\min_i \min_j \Delta_i^- (j) + \zeta \max_i \max_j \Delta_i^- (j)}{\Delta_i^- (j) + \zeta \max_i \max_j \Delta_i^- (j)} \tag{11}$$

where $\Delta_i^+ (j) = \left| A^+ (j) - p_i(j) \right|$ and $\Delta_i^- (j) = \left| A^- (j) - p_i(j) \right|$. $p_i(j)$ can be obtained based on the weighted normalized decision matrix mentioned above. ζ is resolution coefficient and is equal to 0.5.

The Gray correlation coefficient matrix with respect to the positive-ideal solution and negative-ideal solution is thus given by

$$R^+ = \begin{bmatrix} r_{11}^+ & r_{12}^+ & \cdots & r_{1n}^+ \\ r_{21}^+ & r_{22}^+ & \cdots & r_{2n}^+ \\ \cdots & \cdots & \cdots & \cdots \\ r_{m1}^+ & r_{m2}^+ & \cdots & r_{mn}^+ \end{bmatrix}, R^- = \begin{bmatrix} r_{11}^- & r_{12}^- & \cdots & r_{1n}^- \\ r_{21}^- & r_{22}^- & \cdots & r_{2n}^- \\ \cdots & \cdots & \cdots & \cdots \\ r_{m1}^- & r_{m2}^- & \cdots & r_{mn}^- \end{bmatrix} \tag{12}$$

The gray correlation degree of the alternative A_i is shown as following formula

$$R_i^* = \frac{R_i^+}{R_i^+ + R_i^-} \tag{13}$$

where R_i^+ and R_i^- are the grey similarity degree, and can be obtained by

$$R_i^+ = \frac{1}{n} \sum_{j=1}^{n} r_{ij}^+, R_i^- = \frac{1}{n} \sum_{j=1}^{n} r_{ij}^-, i = 1, 2, \cdots, m, j = 1, 2, \cdots, n. \tag{14}$$

Once the gray correlation degrees are determined, the ranking order of all alternatives can be obtained based on the combination of gray correlation degree and the Euclidean distances in TOPSIS. The comprehensive correlation degree respect to the positive and negative ideal solution can be defined as follow

$$S_i^+ = \alpha_1 D_i^- + \alpha_2 R_i^+, S_i^- = \alpha_1 D_i^+ + \alpha_2 R_i^- \tag{15}$$

where $\alpha_1 + \alpha_2 = 1$. α_1 and α_2 can be determined according to the preference of decision makers. The final risk priority index of all the alternatives can be calculated by

$$RI_i = \frac{S_i^+}{S_i^+ + S_i^-}, i = 1, 2, \cdots, m \tag{16}$$

The value of RI_i lies in the interval of (0, 1), which implies that the alternative (A_i) is closer to the positive ideal solution (A) and have the more similar geological relationship with A^+ as the value of RI_i approaches 1. Therefore, using the risk priority index, we can easily determine the risk rank order of all the hydropower dams and select from them the highest priority one.

3 EMPIRICAL STUDY AND RESULTS

A case study of comparing five hydropower dams (A_1, A_2, ..., A_5) in China was conducted to examine the applicability of this risk informed decision making method. Five risk factors (deficiency degree (C_1), management level (C_2), storage capacity (C_3), dam height (C_4), and risk population (C_5)) were identified as the factor criteria for the dam risk prioritization. With the AHP method, the deficiency degree criterion has the highest degree of importance in assessing the overall risk performance of hydropower dams. The management level, storage capacity, dam height and the risk population criterion have a relative low degree of importance for the dams considered. The weight vector of the risk factor is determined as $W = (0.4, 0.15, 0.15, 0.15, 0.15)$.

The description and rating of each criteria are listed in Table 1. The deficiency degree and the management level of the hydropower dam are determined based on the conclusions that are already available in the safety assessment and management assessment of registered hydropower dams. The risk population is obtained by means of the dam break flood analysis and flood inundation map in the downstream area.

By using the available risk factor data of these dams, the performance rating of each dam with respect to each risk factor was calculated. Table 2 shows the result.

To apply the proposed approach, the decision matrix contained in Table 2 needs to be normalized and weighted by Eqs. (2), (3) and (4). Table 3 shows the result. The positive ideal solution and the negative ideal solution are then determined by Eqs. (5) and (6) as $A^+ = (0.4, 0.15, 0.15, 0.15, 0.15)$, $A^- = (0, 0, 0, 0, 0)$.

With the data in Table 3, the Euclidean distances in TOPSIS (D^- and D^+), the grey similarity degree (R^+ and R^-) and the comprehensive correlation degree (S^+ and S^-) of each dam can be calculated. Both α_1 and α_2 are assumed to be equal to 0.5 in the comprehensive correlation degree. Table 4 shows the results. Table 5 demonstrates the relative closeness degree in TOPSIS (D^*), gray correlation degree (R^*) and risk priority index (RI) obtained by Eqs. (9), (13) and (16) and its corresponding ranking.

The results in Table 5 indicate that the risk priority index proposed in this paper has the same ranking with the result of TOPSIS, and it is similar with that of gray correlation degree. That is because the risk priority index is a comprehensive correlation degree which is proposed based on TOPSIS and the gray correlation degree. It reflects not only the closeness of alternative to the ideal solution but also the geometrical relationship. As the modification of α_1 and α_2, the results will reflect the different preference of the relative closeness and the geometrical similarity by the decision makers.

Table 1. Risk prioritization criteria and ratings.

Deficiency Degree	Description	Good	Fair	Poor	Critical		
	Rating	1	2	3	4		
Management Level	Description	Good	Fair	Poor			
	Rating	1	2	3			
Storage capacity /10^8 m^3	Description	<0.01	0.01~0.1	0.1~1	1~10	10~100	≥100
	Rating	1	2	3	4	5	6
Dam height /m	Description	<30	30~50	50~100	100~200	≥200	
	Rating	1	2	3	4	5	
Risk population /people	Description	1~10^2	10^2~10^4	10^4~10^6	>10^6		
	Rating	1	2	3	4		

Table 2. Performance ratings of dams.

	Deficiency degree (C_1)	Management level (C_2)	Storage capacity (C_3)	Dam height (C_4)	Risk population (C_5)
A1	4	1	3	3	3
A2	3	1	5	1	3
A3	1	1	6	5	4
A4	1	2	3	2	3
A5	1	3	1	1	1

Table 3. Weighted normalized decision matrix.

	Deficiency degree (C_1)	Management level (C_2)	Storage capacity (C_3)	Dam height (C_4)	Risk population (C_5)
A1	0.4	0	0.06	0.075	0.1
A2	0.2667	0	0.12	0	0.1
A3	0	0	0.15	0.15	0.15
A4	0	0.075	0.06	0.0375	0.1
A5	0	0.15	0	0	0

83

Table 4. Performance comparison of dams.

	D^-	D^+	R^+	R^-	S^+	S^-
A1	0.423	0.197	0.758	0.699	0.591	0.448
A2	0.309	0.257	0.682	0.744	0.496	0.501
A3	0.260	0.427	0.781	0.743	0.520	0.585
A4	0.144	0.435	0.638	0.801	0.391	0.618
A5	0.150	0.477	0.610	0.914	0.380	0.696

Table 5. Risk performance index and ranking of dams.

	TOPSIS D^*	Gray correlation degree R^*	Risk priority index RI	Ranking of D^*	Ranking of R^*	Ranking of RI
A1	0.683	0.520	0.569	1	1	1
A2	0.546	0.478	0.498	2	3	2
A3	0.378	0.513	0.471	3	2	3
A4	0.248	0.443	0.387	4	4	4
A5	0.239	0.400	0.353	5	5	5

4 CONCLUSION

The new decision making method described in this paper provides an effective framework for ranking hydropower dams in terms of their overall risk performance in a simple and transparent manner using existing multi-criteria decision tools and strategies. This method combines the TOPSIS method and the gray correlation degree method to ensure a meaningful interpretation of the comparison result.

The empirical study of real cases in China demonstrates that the method can effectively reflect the decision information emitted by the risk factors, and provide meaningful rankings and useful information. The method is computationally simple and its underlying concept is rational and comprehensible, thus facilitating its implementation in a computer-based system.

ACKNOWLEDGEMENT

This work was supported by the Postdoctoral research funding project of Zhejiang Province, China (BSH1502042).

REFERENCES

Andersen G R, Chouinard L E, Hover W H, et al. Risk indexing tool to assist in prioritizing improvements to embankment dam inventories[J]. Journal of geotechnical and geoenvironmental engineering, 2001,127(4):325–334.

Hartford D N D. Lessons from 20+ Years of Experience and Future Directions of Risk Informed Dam Safety Management: 82nd Annual Meeting of ICOLD, Bali, Indonesia, 2014[C].

Hwang, C.L., Yoon, K. 1981. Multiple Attribute Decision Making: Methods and Applications, Spring-Verlag, New York.

Lai, Y.J., Liu, T.Y., Hwang, C.L. 1994. TOPSIS for MODM, *European journal of Operational Research* 76(3): 486–500.

Saaty, T.L. 1995. Decision making for leaders. 3rd ed. *New York: RWS Publications.*

Yoon, K., Hwang, C.L. 1995. *Multiple Attribute Decision Making: An Introduction.* Sage, Thousand Oaks, CA.

Hydraulic Engineering IV – Xie (Ed.)
© 2016 Taylor & Francis Group, London, ISBN 978-1-138-02948-4

Initial exploration of the application of steel reinforced concrete in radial gate pier

Xiaofei Zhang, Yaling Liu, Kaiman Wei, Ruichang Hu, Qingfu Zeng & Chun Yuan
Guangxi University, Nanning, Guangxi, China

ABSTRACT: To strengthen the high thrust radial gate pier, a new type of steel reinforced concrete pier structure is proposed in this paper. The shape steel is used to replace the fan-shaped steel reinforcement in the conventional reinforced concrete pier. By using finite element software, a 3D nonlinear finite element model is developed to analyze the performance of the steel reinforced concrete pier with respect to deformation, stress, crack development, and bearing capacity of the pier. It can be concluded that the performance of the steel reinforced concrete pier is similar to that of the conventional reinforced concrete pier. But the bearing capacity and ductility of the reinforced concrete pier are significantly improved, which shows that it is worth further studying the application of the steel reinforced concrete structure in the radial gate pier.

1 INTRODUCTION

With the increasing discharge of the water release structure in large-scale water projects, as well as the working head and the expanding of the aperture dimension, the thrust to the radial gate pier is also enhanced. It is more difficult for the conventional reinforced concrete pier to satisfy the requirements of a reliable operation. Current methods that are used to solve this problem is usually pre-stressed technology which is applied to the pier structure (Gao & Xu, 2014), and the stress state of the pre-stressed pier is relieved effectively, but it is generally prone to have a large pull-anchor coefficient. In addition, the design and construction of the pre-stressed pier is quite complex (Xie & Li, 2010).

This paper is based on a proposed design of the radial sluice gate pier of Yongning Hydraulic Complex in Nanning, Guangxi. Given that the steel reinforced concrete structure has the characteristics of high bearing capacity, large stiffness, and good ductility (Xue, 2010, Zhao, 2001), it is appropriate to be applied to the radial gate pier. In this paper, we proposed a novel structure model of the steel concrete pier. Certain procedures can be performed to replace the fan-shaped bearing force reinforcement in the pier body and the longitudinal reinforcement in the bracket, which are widely used in the conventional reinforced concrete pier, with the shape steel. Compared with the conventional reinforced concrete pier, the bearing capacity of the reinforced concrete pier is greatly improved and is more convenient for construction.

2 ARRANGEMENT FORM OF THE STEEL REINFORCED CONCRETE STRUCTURE PIER

In the context of engineering practice, this paper studies the engineering background of the radial floodgate pier of Yongning Hydraulic Complex in Nanning, Guangxi. The discharging sluice dam of the project has 13 overflow gate holes, and the main gate is the radial gate, the sluice opening size is 20×12 m (width $\times$ height), the thickness of the middle pier is 3.0 m, the overflow weir of sluice openings from no. 1 to no. 10 uses WES Weir, the weir crest elevation is 55.00 m, and transverse joints are designed in the middle of the weir.

One of the design conditions is the normal operation condition when both sides of the gate are closed; the load acting on the radial gate bracket is given in Table 1.

In order to compare the performance of the steel reinforced concrete pier with the conventional reinforced concrete pier, as well as to demonstrate whether the steel reinforced concrete pier is reasonable, on the basis of the "Specification for Hydraulic Reinforced Concrete Structures" (SL191-2008) and the "Handbook of Hydraulic Structure Design" (2013 edition), the conventional reinforced concrete pier is designed. Fan-shaped tensile reinforcements are configured in the body of the pier with an arrangement of a long and short spacing. There are 13 beams of steel reinforcements, and each beam has an included angle of 5 degrees, the length of the long steel reinforcement is 20 m, and the short one is 15 m. The longitudinal bearing force reinforcement, which is located in the bracket, is positioned inside the whole pier, thus stretching from the lateral bracket to the other side of the outside bracket. The calculation of the section reinforcement quantity is based on the formula proposed in the "Handbook of Hydraulic Structure Design" (2013 edition) (Zeng, 2014, Yuan, 2014).

The area of the fan-shaped reinforcement in the tensile region of the pier is given as follows:

$$F \le \frac{1}{\gamma_d} f_y \sum_{i=1}^{n} A_{si} \cos \theta_i \qquad (1)$$

The area of the longitudinal reinforcement in the bracket is given as follows:

$$A_S = \frac{\gamma_d Fa}{0.8 f_y h_0}. \qquad (2)$$

In the conventional concrete pier, the fan-shaped bearing force reinforcement in the pier body and the longitudinal reinforcement in brackets are widely used. In this paper, we used the shape steel in the pier body and brackets to replace them. Shape steel in the pier body is arranged along the direction of the resultant force, which is mainly applied in the gate thrust. The shape steel in the bracket passes through the bracket and vertically connects with the shape steel in the pier. The steel reinforced concrete pier has the same quantity of steel as it is in conventional reinforced concrete pier. The length of shape steel in pier body is 20 m, and the strength of shape steel in steel reinforced concrete pier and the reinforcement in the conventional concrete pier is consistent. The section of the shape steel is H-section steel. The arrangement forms of the conventional reinforced concrete gate pier and the steel reinforced concrete pier are shown in Figures 1 and 2, respectively.

3   THE FINITE ELEMENT MODEL OF THE PIER

The unit division diagrams of the conventional reinforced concrete pier and the steel reinforced concrete radial gate pier are shown in Figures 3 and 4, respectively. The element type of concrete chooses SOLID65 element, and its failure criterion adopts Willam–Warnke's five-parameter criterion. The element type of the steel reinforcement adopts Link8 element, and the multiple linear with a kinematic hardening model is selected as the constitutive model. The shape steel used Beam188 element for simulation, and multiple linear with a

Table 1.   Loads acting on the radical gate bracket.

Condition	X (KN)	Y (KN)	Z (KN)
Normal operation condition	7400	5020	1500

*The positive direction of X is vertical to the axis of the dam and downstream, the positive direction of Y is vertically upward, and the positive direction of Z is parallel to the dam axis to the right bank.

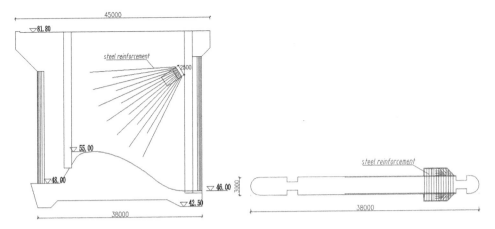

Figure 1.   Arrangement form of the conventional reinforced concrete pier.

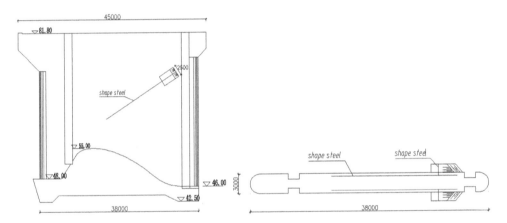

Figure 2.   Arrangement form of the steel reinforced concrete pier.

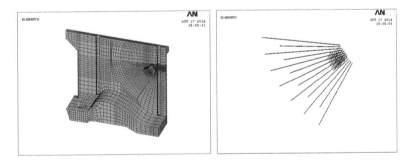

Figure 3.   Finite element model of the conventional reinforced concrete pier.

kinematic hardening model is considered as the constitutive model. The separated model is used to build the model of the steel reinforcement and concrete. Rigid connection is considered as the connection style between shape steel and concrete, and the effects of bond slip is not considered.

87

4 COMPARATIVE ANALYSIS OF THE PERFORMANCE OF THE STEEL REINFORCED CONCRETE RADIAL GATE PIER AND THE CONVENTIONAL REINFORCED CONCRETE PIER

4.1 *Analysis of displacement at the center of the bracket*

The deformation of the bracket has a direct impact on the normal operation of the gate and the safety of the pier, as it is located directly under the load. Under the design load (under the conditions that both sides of the gate are closed), the results of the finite element calculation for deformation at the center of the bracket in the steel reinforced concrete radial gate pier and the conventional reinforced concrete pier are summarized in Tables 2 and 3.

From Tables 2 and 3, we can draw the conclusion that the displacement at the center of the bracket between two pier structures has little difference, except in the ultimate state, and the displacement differences between two pier structures under 2 and 3 times the design load are 0.4 and 17 mm, respectively, and in the ultimate state, it becomes 92 mm. The results indicate that the steel reinforced concrete pier has obviously improved in terms of ultimate bearing capacity and ductility compared with the conventional reinforced concrete pier.

4.2 *Stress analysis of the pier*

From the stress distribution of the pier, it can be seen that there is no significant difference between the steel reinforced concrete pier and the conventional reinforced concrete

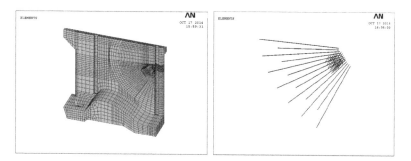

Figure 4.   Finite element model of the steel reinforced concrete pier.

Table 2.   Displacement data tables of the steel reinforced concrete pier at the middle of the loading surface of the bracket under different loads.

Load	X (m)	Y (m)	Z (m)	USUM (m)
Design load	0.001	0.000	0.000	0.001
2 times the design load	0.008	0.004	0.003	0.010
3 times the design load	0.036	0.028	0.011	0.047
Ultimate load (5.17 times the design load)	0.135	0.098	0.035	0.171

Table 3.   Displacement data tables of the conventional reinforced concrete pier at the middle of the loading surface of the bracket under different load.

Load	X (m)	Y (m)	Z (m)	USUM (m)
Design load	0.001	0.000	0.000	0.001
2 times the design load	0.006	0.003	0.002	0.006
3 times the design load	0.025	0.015	0.007	0.030
Ultimate load (3.92 times the design load)	0.062	0.045	0.019	0.079

pier in stress distribution disciplinarian. The bracket deforms along the loading direction when subjected to loads, causing a large tensile stress in concrete around the bracket and the downstream maintenance gate slot; the compressive stress mainly appears in the bracket and concrete near the interface of the bracket and the pier. Under the design load, the maximum tensile stress values in the steel reinforced concrete pier and the conventional reinforced concrete pier are 1.25 Mpa and 1.31 MPa, respectively. Besides, the maximum tensile stress works in the middle of the upper surfaces, which is located in the bracket and the pier. The stress distribution of the steel reinforced concrete pier and the conventional reinforced concrete pier under the design load is shown in Figures 5 and 6.

4.3 The comparative analysis of the development of the cracks in the pier concrete

Judging from the developing process of cracks in the pier concrete, the development and distribution of the cracks in the steel reinforced concrete pier are similar to those of the conventional reinforced concrete pier under the normal operating condition. In the initial stage, the load is relatively small, and the stress of the whole structure is also relatively small at this time. Therefore, the concrete has not yet begun to crack. When the load reaches 0.67 times the design load, the concrete of two pier structures begins to crack, the cracks developed along the interface between the bracket and the pier. At this time, there is no significant difference in crack distribution between two pier structures. Cracks began to develop around the interface of the bracket and the pier if the load is increased. When the existing load reaches to the designed load, the cracks begin to develop towards the downstream side of the interface, but

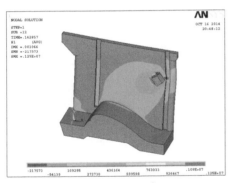

(a) Steel reinforced concrete pier.

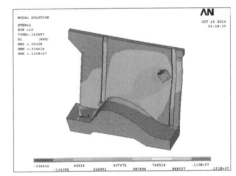

(b) Conventional reinforced concrete pier.

Figure 5.   The first principal stress distribution of the pier (Pa).

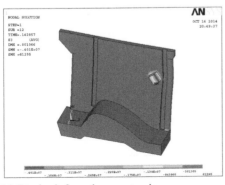

(a) Steel reinforced concrete pier.

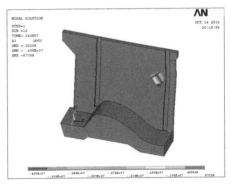

(b) Conventional reinforced concrete pier.

Figure 6.   The third principal stress distribution of the pier (Pa).

there are no obvious vertical cracks in the pier. If the load continues to increase, the crack will further expand all around. When the load is two times the design load, cracks of the steel reinforced concrete pier develop faster than those of the conventional reinforced concrete pier, and the crack distribution area is higher than that of the conventional reinforced concrete pier. The cracks begins to develop rapidly if the load is increased. When the load reaches the three times the designed load, the crack distribution area of the steel reinforced concrete pier is higher than that of the conventional reinforced concrete. When the pier reaches its ultimate bearing capacity, most of the concrete cracks. The crack distributions of two pier structures under the condition of two times the design load are shown in Figure 7.

4.4 *Comparative analysis of the carrying capacity*

As shown in Table 1, the load of the normal operating conditions is used as the base load. Load in all directions was equal when the load increased proportionally. The load–displacement curves of the steel reinforced concrete pier and the conventional reinforced concrete pier under various load levels are shown in Figure 8.

From Figure 8, we can see that under the monotonic loading condition, the development of displacement between the steel reinforced concrete pier and the conventional reinforced concrete pier is the same. At the initial stage of loading, the structure is in the elastic state and there is no significant difference in displacement. When the load is increased to a certain extent, the concrete begins to crack, but it still works together with the reinforcement or the shape steel to bear the increasing load. Due to the fan-shaped distribution of the steel, the transmission range of force is comparatively wide, which results in a relatively small displace-

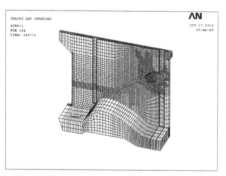 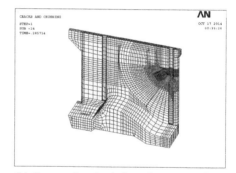

(a)Steel reinforced concrete pier.   (b) Conventional reinforced concrete pier.

Figure 7.   The crack distribution of the pier.

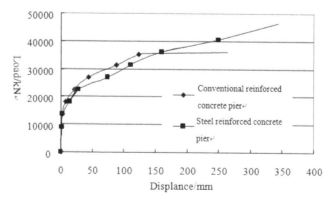

Figure 8.   Load–displacement curves.

Table 4. The data of the ultimate load and ultimate displacement of piers.

Models	Steel reinforced concrete pier	Conventional reinforced concrete pier
Design load (kN)	9067	9067
Ultimate load (kN)	46,530	35,280
Ultimate displacement (mm)	342	123

ment of the conventional reinforced concrete pier; if the load is increased continuously, the stress of the reinforcement and the shape steel increases accordingly and enters the yield state; when the load reaches a certain level, the partial reinforcement is ineffective due to the yielding. However, the shape steel can still work together with the concrete to bear the load. The ultimate load of the steel reinforced concrete damage is relatively large. The ultimate load and the ultimate displacement results of the pier when the structure is damaged are summarized in Table 4.

5 CONCLUSION

1. The displacement and stress distribution of concrete in the steel reinforced concrete pier and the conventional reinforced concrete gate pier are generally the same. Besides, there are no obvious differences in displacement at the center of the bracket between two pier structures, except in the ultimate state, and the displacement of two pier structures is basically the same under the designed load level.
2. When conducting research on the load action of the sluice gate on both sides under the closed state, the development and distribution of cracks in the steel reinforced concrete piers are found to be similar to that of the conventional reinforced concrete pier. The initial crack load is the same; with the increasing load, the range of concrete cracks in the steel reinforced concrete pier becomes slightly larger than that in the conventional reinforced concrete pier. In addition, the crack resistance performance of the steel reinforced concrete pier is not as good as the conventional reinforced concrete pier.
3. The load is considered as the base load when the sluice gate on both sides is closed. The ratio of the ultimate load and the base load is 5.17 in the steel reinforced concrete radial gate pier and 3.92 in the steel reinforced concrete pier. The ultimate bearing capacity of the steel reinforced concrete pier is better than that of the conventional steel high concrete pier.
4. The performance of the steel reinforced concrete pier is similar to that of the conventional reinforced concrete pier, and the bearing capacity and the ductility of the steel reinforced concrete pier are obviously enhanced.
5. The steel occupies a very small portion of the pier body and is convenient to construct. It is conducive to the application of the steel reinforced concrete pier in the large-thrust gate piers. Compared with the conventional reinforced concrete pier, the steel reinforced concrete pier possesses a more significant increase in ultimate bearing capacity as well as ductility. However, it exhibits poor performance in terms of the limited crack after subjected to the same amount of reinforcement. With the development and application of the high performance concrete in tensile, crack, such as the fiber one, the limited crack foundation of the sluice pier can match well with the bearing capacity. Therefore, applying the steel reinforced concrete pier to the large-thrust radial gate pier is feasible and potential, which is worthy of further exploration.

REFERENCES

Chun Yuan. 2014. The impact studies of steel sectional form on steel reinforced concrete pier carrying capacity. Nanning: Guangxi University.

Hongtie Zhao. 2001. *Steel-concrete composite structures.* Beijing: Science Press.

Jianyang Xue, 2010. Design principles of steel and concrete composite structures. Beijing: Science Press.

Kaixu Gao & Shangwei Xu etc. 2014. Stress analysis and research on pre-stressed piers with steel reinforced concrete bracket. Water Resources Planning and Design.25(7):80–82.

Qingfu Zeng. 2014. The impact studies of protective layer thickness on steel reinforced concrete pier carrying capacity. Nanning: Guangxi University.

Wei Xie & Shushan Li. 2010. *Test and theory of pre-stressed pier structure.* Beijing: China Water Conservancy and Hydropower Publishing House.

Hydraulic Engineering IV – Xie (Ed.)
© 2016 Taylor & Francis Group, London, ISBN 978-1-138-02948-4

Optimal schedule for track maintenance actions for fixed and seasonal possession costs

Mohammad Daddow, Xiedong Zhang & Hongsheng Qiu
Wuhan University of Technology, Wuhan, Hubei, China

ABSTRACT: A mathematical model for the track maintenance scheduling problem is presented in this paper. The model improves a previously proposed one by using seasonal track possession costs. A more general cost structure is reported: a structure has the possibility of including penalty cost paid if the first execution for each maintenance action is planned too early in the first allowed cycle compared with its end. The aim of this study is to get the optimal schedule for those actions over one segment of track in a finite horizon, and to discuss the influence of a set of non-combinable works and possession costs on the optimal solution. The model was tested for two scenarios by using MATLAB R2014b. In conclusion, it was found that the set of non-combinable works and the possession costs play important roles in the optimal schedule and that the model preferred to avoid planning in the periods concurred with months of high possession costs as much as possible.

1 INTRODUCTION

Maintenance work for railway tracks are very costly, therefore, efficient maintenance plans and optimal schedules for them are needed in order to reduce the total cost without reducing the maintenance itself (Soh *et al.* 2012). Besides, the Track Maintenance Scheduling Problem (TMSP) of the extensive railway networks is a very complex one. Usually, the strategy of maintenance decisions for the tracks has been based on previous experience and judgment of experts. Nowadays, there is substantial labor involved in order to improve this strategy by using operation research methods to get a better solution to this problem (Patra 2009, Peng 2011). This research considers only the preventive maintenance, and many papers have focused on the Preventive Maintenance Scheduling Problem (PMSP) (Canfield 1986, Higgins 1998, Dahal & Chakpitak 2007, Canto 2008, Peng *et al.* 2011, Moghaddam & Usher 2011, Gustavsson *et al.* 2014).

The PMSP and restricted Preventive Maintenance Scheduling Problem (RPMSP) models for railway infrastructures with respect to the concept of track possession costs have only been studied in recent years by Budai *et al.* (2006), in which routine activities and project works were formulated differently to minimize the sum of the possession costs and the maintenance costs for one link of the railway track. The possession cost is defined by the possession time, the time that a track needs for maintenance works and cannot be used by trains. Moreover, decisionmakers assumed that the possession cost is constant for each period within the planning horizon and solved these models using four heuristics. Thereafter, Pouryousef *et al.* (2010) further upgraded the RPMSP model for dedicated and mixed high-speed rail scenarios for many segments, enforcing tough frequencies on routine activities and handling traffic restrictions with cost penalties for track possession during specific periods within the planning horizon.

These studies are very valuable and could be applied to a large group of applications; however, to the best of our knowledge, no analysis is yet available on TMSP when the track possession cost varies from one season to another within the planning horizon. Likewise, the comparison between its optimal solutions for fixed and seasonal track possession costs under different scenarios for a set of non-combinable works has also not been investigated. The aim of this research is to present an extension of the mathematical model developed by Pouryousef *et al.* (2010) to get an optimal solution for a given set of maintenance actions for one segment of track in a limited horizon under the above circumstances, minimizing the overall cost. In addition, the optimal solutions obtained are analyzed in order to understand the influence of track possession costs on the optimal schedule of the studied problem.

2 DESCRIPTION OF THE PROBLEM

The preventive maintenance actions on a track can be classified into two basic types: 1) small routine actions, which consist of inspections and small repairs (e.g. inspection of level crossings and lubrication of switches) that are done regularly with frequencies varying between monthly and annual; and 2) projects, which include longer renewal works (e.g. ballast cleaning) that are performed only once or twice every few years (Budai *et al.* 2006).

The problem output is to get an optimal schedule for those actions in the planning horizon, where actions can be planned as much as possible in the same period by minimizing the total cost.

The required problem input in this study can be briefly described as follows:

1. A set of routine actions and projects is given, and then it is required to plan them in a suitable schedule. Some routine works and projects may be combined to reduce the possession costs, but others may exclude each other.
2. The planning cycle length for each routine maintenance work (i.e. the maximum number of time periods between two successive executions) is known and given in advance.
3. The number of time periods elapses for each routine action, as it has been performed for the last time before the start of the planning horizon.
4. The maintenance costs of each routine action and project and the possession cost per period.
5. A set of non-combinable works (i.e. a set of routine works and projects that cannot be planned simultaneously on the same segment).
6. The planning horizon.

Finally, for more facilitation of running the upgraded RPMSP model considered in this research, only routine maintenance works (seven actions) were considered through the scenarios. Therefore, we did not consider the projects during the mathematical formulation.

3 MATHEMATICAL FORMULATION

3.1 *Sets and parameters*

The sets and parameters are given as follows:

- *MA*: set of routine preventive maintenance actions;
- *D*: {(*m,n*)| action *m* cannot be combined with *n*, $\forall m, n \in MA$} set of non-combinable actions;
- p^a: planning cycle length of the routine action $a \in MA$;
- E^a: number of time periods elapsed for routine action $a \in MA$, as it was carried out for the last time before starting the planning horizon;
- AC^a: $(p^a - E^a)$ length of the first allowed planning cycle of the routine action $a \in MA$;
- FC^a: {$t \in T$| $1 \leq t \leq AC^a$}$\subseteq T$ set of time periods of the first allowed planning cycle for the routine action $a \in MA$;

- LC^a: $\{t \in T \mid 1+|T|-p^a \le t \le |T|\} \subseteq T$ set of time periods of the last planning cycle for the routine maintenance action $a \in MA$;
- g_t^a: $(AC^a - t)/AC^a$ the remaining interval length between the time period $t \in FC^a$ and the end of the first allowed planning cycle of the routine action $a \in MA$ divided by the length of this cycle;
- h_t^a: $(|T|-t)/p^a$ the remaining interval length between the time period $t \in LC^a$ and the end of planning horizon divided by the cycle length for the routine action $a \in MA$;
- Se: $\{Wi, Sp, Su, Au\}$ set of year seasons;
- pc_t^{se}: possession cost in the period $t \in T$ over each month of a given season $se \in Se$;
- Mc_a: maintenance cost per period for carrying out the maintenance action $a \in MA$;

3.2 Binary decision variables

The binary decision variables are given as follows:

- u_t^a: binary decision variable that indicates whether the maintenance action $a \in MA$ is planned in the period $t \in T$ ($u_t^a = 1$) or not ($u_t^a = 0$);
- v_t: binary decision variable that indicates whether the track is used for the maintenance at the period $t \in T$ ($v_t = 1$) or not ($v_t = 0$).

3.3 Model

The model is given as follows:

$$Min \sum_{se \in Se} \sum_{t \in T} pc_t^{se} v_t + \sum_{a \in MA} \sum_{t \in T} Mc_a u_t^a + \sum_{a \in MA} \sum_{t \in FC^a} Mc_a g_t^a u_t^a + \sum_{a \in MA} \sum_{t \in LC^a} Mc_a h_t^a u_t^a \qquad (1)$$

which is subjected to:

$$\sum_{t=1}^{AC^a} u_t^a = 1 \, (a \in MA), \qquad (2)$$

$$u_t^a = u_{t+qp^a}^a \, (a \in MA, 1 \le t \le |T| - P^a, q \ge 1), \qquad (3)$$

$$u_t^m + u_t^n \le 1 \, (t \in T, (m,n) \in D), \qquad (4)$$

$$v_t \ge u_t^a \, (a \in MA, t \in T), \qquad (5)$$

$$u_t^a, v_t \in \{0,1\} \, (a \in MA, t \in T). \qquad (6)$$

The objective function in Equation 1 minimizes the sum of possession costs, the maintenance costs, the penalty cost paid if the first execution of the routine actions is planned too early in the first allowed planning cycle compared with its end, and the penalty cost considered if the last execution of the routine actions is scheduled too early in the planning horizon compared with the end of horizon. Furthermore, constraints of Equation 2 ensure that each action is planned exactly once in the first allowed planning cycle, and constraints of Equation 3 guarantee that until the end of the planning horizon, the activities for the other cycles will be determined as well, ensuring exactly p^a periods between two consecutive executions of the same activity. While constraints of Equation 4 guarantee that non-combinable works cannot be planned simultaneously on the same segment, and constraints of Equation 5 ensure that the period $t \in T$ will be assigned for the maintenance action if and only if at least one action is scheduled for that period on this segment. Finally, constraints of Equation 6 guarantee that all decision variables are binary.

4 COMPUTATIONAL EXPERIMENTS

4.1 *General assumptions and requirements*

The general assumptions and requirements are as follows:

1. Instances were carried out on a personal computer with an Intel i5 processor operating at 2.40 GHz and RAM of 4 GB. Also, an algorithm was developed in MATLAB R2014b software to get a solution for the model formulated in Equations 1–6.
2. Only routine maintenance works (seven works) were considered for running the model over the suggested scenarios, and the planning horizon was started on 1stDecember.
3. $E^a = 0$, $\forall a \in MA$: in order to increase the number of possible schedules for each action.
4. It can choose for each action any period in the first cycle to schedule it, but for the other cycles and until the end of planning horizon, the duration between the two consecutive executions is restricted and equal to the cycle length. For instance, Table 1listsall possible schedules for routine work (*a*) with a cycle length of 3 months in the planning horizon of 1 year.

4.2 *Related data*

It was assumed that the planning horizon for all instances is 2 years and the discrete time periods are months. Also, we assumed that the track possession cost has two patterns:

1. Fixed possession (first pattern): the possession cost value is the same for each month within the planning horizon.
2. Seasonal possession (second pattern): the possession cost value is different for each season during the years of the planning horizon.

To evaluate the suggested model, researchers took a set of seven routine actions; their cycle lengths and maintenance costs per month are presented in Table 2.

The model was tested for two scenarios. In the first scenario (scenario (A)), it was supposed that each routine action can be combined with all other routine actions at the same time (i.e. $D_A = \varnothing$). In the second scenario (scenario (B)), it was assumed that the set D consists of two subsets of non-combinable works: (1) a subset of two works and (2) a subset of three works (e.g. $D_B = \{(r_1, r_5), (r_2, r_6, r_7)\}$). Besides, the model was tested for the two previous patterns, where the monthly costs of track possession for each season are presented in Table 3. In this table, it was considered that all possible cases for the second pattern from S1Pc to S6Pc were based on those for winter months only. First, the costs for winter months are

Table 1. All possible schedules for routine work (*a*).

Possible schedules	Discrete time periods (month)											
	1	2	3	4	5	6	7	8	9	10	11	12
1	1	0	0	1	0	0	1	0	0	1	0	0
2	0	1	0	0	1	0	0	1	0	0	1	0
3	0	0	1	0	0	1	0	0	1	0	0	1

Table 2. Routine actions and their cycle lengths and maintenance costs per month.

Routine action	r_1	r_2	r_3	r_4	r_5	r_6	r_7
Cycle length (month)*	3	3	3	4	6	6	18
Maintenance cost (10^3 USD)	4	5	6	8	10	11	24

*Data for cycle lengths from Pouryousef *et al.* (2010).

Table 3. Monthly costs of track possession for each season.

Possession cost pattern	Encoding*	Winter (10³ USD)	Spring (10³ USD)	Summer (10³ USD)	Autumn (10³ USD)
Fixed	FPc	50	50	50	50
Seasonal	S1Pc	75	50	50	50
	S2Pc	75	50	50	25
	S3Pc	100	75	50	25
	S4Pc	75	75	50	50
	S5Pc	75	75	50	25
	S6Pc	75	75	75	50

*FPc = fixed cost, S1Pc to S6Pc = studied cases of seasonal cost.

Table 4. Computational results for scenarios (A) and (B).

Possession cost cases	Optimal solutions*		Optimal costs (10³ USD)	
	Scenario (A)	Scenario (B)	Scenario (A)	Scenario (B)
FPc	A-1	B-1	884	999.3
S1Pc	A-1	B-1	934	1049.3
S2Pc	A-2	B-2	838	923.3
S3Pc	A-2	B-2	938	1073.3
S4Pc	A-2	B-2	988	1173.3
S5Pc	A-2	B-2	888	1023.3
S6Pc	A-2	B-2	1088	1223.3

*Sets of non-combinable works: $D_A = \varnothing$ and $D_B = \{(r_1, r_5), (r_2, r_6, r_7)\}$.

the highest, those for other seasons' months are the same for S1Pc and those for spring and summer months are the same only for S2Pc but different for S3Pc. Second, the costs for winter and spring months are the same and highest while those for other two seasons' months are the same for S4Pc but different for S5Pc. Finally, all costs are the same except for those in autumn months for S6Pc.

4.3 Results for scenarios (A) and (B)

Test results for the instances of these two scenarios are summarized in Table 4, and the optimal schedules for each scenario are shown in Figure 1. In each scenario, it was found that only two optimal solutions were distributed between the studied cases of the track possession cost, which are A-1 and A-2 for scenario (A) and B-1and B-2 for scenario (B) (see Figure 1). Moreover, it can be noted from these solutions that the actions, which have the same cycle lengths (e.g. r_1, r_2 and r_3) and can perform them together, were planned in the same periods.

In scenario (A), the only difference between its two schedules is that the plan of the action r_4 in A-1 (related to fixed possession costs (FPc) and S1Pc) starts in the fourth period while it starts in the third period in A-2 (related to S2Pc–S6Pc). This plan (in each of the two schedules)created four additional possession periods on the other actions, which are 4, 8, 16, and 20 (which fall in spring and summer months) in A-1 and 7, 11, 19, and 23 (which fall in summer and autumn months) in A-2.ForFPc and S1Pc in both plans of the action r_4, it can be observed that the summation of possession costs for those additional periods is the same and equal to 200, but the model chose for them the schedule that starts in the fourth period (A-1),which gives the minimum penalties for the planning of the first and last cycles of that action. While for S2Pc–S6Pc, the model chose the beginning of the action r_4 in the third period (A-2) in order to take the lowest values for possession costs, which fall in summer and

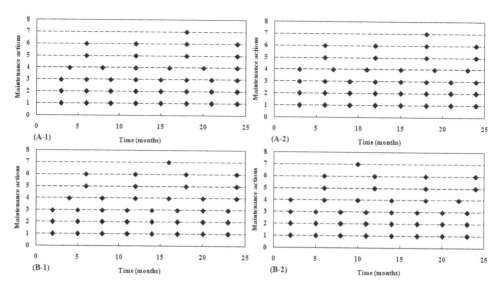

Figure 1.   Optimal schedules in scenarios (A) and (B).

Season	Autumn			Winter			Spring		
Possession cost	25	25	25	⟨100⟩	100	⟨100⟩	⟨75⟩	75	75
Periods	[10]	11	12	13	14	15	16	17	18
Planned actions	r_4, r_7	⟨r_2⟩	⟨r_6⟩	✗	⟨r_2⟩	✗	✗	⟨r_2⟩	⟨r_6⟩

⇧

Figure 2.   Illustration for scheduling the action r_7 in the solution B-2.

autumn months. This situation indicates that the model prefers to avoid planning in the periods of time concurred with months of high track possession costs in the seasonal possession costs as much as possible.

In scenario (B), the differences between its two solutions are that in B-1, the actions r_4 and r_7 start in the fourth and sixteenth periods, respectively, while in B-2, they start in the second and tenth ones, respectively. These differences were caused by the influence of set D, the penalty costs, and the cases of the track possession cost. For instance, in the B-1 solution, the action r_7 was not planned in the last possible period (i.e. the eighteenth one) because the action r_6 was planned during that period. Also, it was not planned in the seventeenth period because the action r_2 was planned in that period, where the set D contains the subset of (r_2, r_6, r_7). Therefore, the model planned the same action in the sixteenth period (its possession cost is equal to 50) to ensure both conditions of set D, thus minimizing the penalty costs related to the early planning in its first and last cycles. While in the B-2 solution, the action r_7 was planned in the tenth period with the action r_4 because the periods between the eighteenth and eleventh ones have either actions that cannot plan with it or high possession costs, as shown in Figure 2.

Finally, the optimal schedules in the two scenarios related to FPc are mostly different from those for the cases of seasonal costs except for S1Pc, where its values are the same as FPc in all seasons except the winter season. Besides, it can be seen in Figure 1 that although the total number of planned actions in both scenarios is equal to 39, the number of possession periods through scenario (A) is equal to 12, while it increases in scenario (B) to become 14 due to the exclusion between the non-combinable works of set D in the last scenario. This leads to suggest that the values of the optimal costs in scenario (A) for each case of track possession are less than those in scenario (B), as shown in Figure 3.

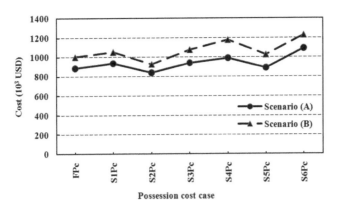

Figure 3. Comparison between the scenarios' optimal costs.

5 CONCLUSION AND RECOMMENDATIONS

In this research, the goal is to get an optimal solution for the **PMSP** for a set of maintenance actions over one segment of track in a limited horizon, minimizing the overall cost, as well as to discuss the influence of the track possession cost and a set of non-combinable works on the final schedule for this problem.

From the computational results, the following conclusion can be drawn:

1. The actions having the same cycle lengths and performing them together were planned in the same periods.
2. The model preferred to avoid planning in the periods of time concurred with months of high possession costs in the seasonal possession cases as much as possible.
3. The values of the optimal costs in scenario (A) were less than those in scenario (B) because of the exclusion between the non-combinable works of set D in the last scenario.

However, the suggested model can be extended to take into account many parameters and constraints that were not considered in this research, such as available workforce, given budget, and multi segments of track.

REFERENCES

Budai, G. Huisman, D. Dekker, R. 2006. Scheduling preventive railway maintenance activities. *Journal of the Operational Research Society* 57(9): 1035–1044.

Canfield, R.V. 1986. Cost optimization of periodic preventive maintenance. *IEEE Transactions on Reliability* 35(1): 78–81.

Canto, S.P. 2008. Application of Benders' decomposition to power plant preventive maintenance scheduling. *European Journal of Operational Research* 184(2): 759–777.

Dahal, K.P. Chakpitak, N. 2007. Generator maintenance scheduling in power systems using metaheuristic-based hybrid approaches. *Electric Power Systems Research* 77(7): 771–779.

Gustavsson, E. Patriksson, M. Strömberg, A.B. Wojciechowski, A. Önnheim, M. 2014. Preventive maintenance scheduling of multi-component systems with interval costs. *Computers & Industrial Engineering* 76: 390–400.

Higgins, A. 1998. Scheduling of railway track maintenance activities and crews. *Journal of the Operational Research Society* 49(10): 1026–1033.

Moghaddam, K.S. Usher, J.S. 2011. Sensitivity analysis and comparison of algorithms in preventive maintenance and replacement scheduling optimization models. *Computers & Industrial Engineering* 61(1): 64–75.

Patra, A.P. 2009. Maintenance decision support models for railway infrastructure using RAMS and LCC analyses. Doctoral thesis, Luleå University of Technology, Luleå, Sweden.

Peng, F. 2011. *Scheduling of track inspection and maintenance activities in railroad networks*. Ph.D. dissertation, University of Illinois at Urbana-Champaign, Urbana, IL.

Peng, F. Kang, S. Li, X. Ouyang, Y. Somani, K. Acharya, D. 2011. A heuristic approach to the railroad track maintenance scheduling problem. *Computer-Aided Civil and Infrastructure Engineering* 26(2): 129–145.

Pouryousef, H. Teixeira, P. Sussman, J. 2010. Track maintenance scheduling and its interactions with operations: Dedicated and mixed high-speed rail (HSR) scenarios. *In Proceedings of the Joint Rail Conference, Urbana, IL, 27–29 April 2010*. American Society of Mechanical Engineers, pp. 317–326.

Soh, S.S. Radzi, N.H. Haron, H. 2012. Review on scheduling techniques of preventive maintenance activities of railway. *In Proceedings of the 4th International Conference on Computational Intelligence, Modelling and Simulation*. IEEE Computer Society, pp. 310–315.

Hydraulic Engineering IV – Xie (Ed.)
© *2016 Taylor & Francis Group, London, ISBN 978-1-138-02948-4*

Effect of bulk solids on strength of cylindrical corrugated steel silos without columns during filling

N. Kuczynska, M. Wojcik & J. Tejchman
Gdansk University of Technology, Poland

ABSTRACT: The paper presents 3D numerical analysis results on the effect of bulk solid on strength and stability of metal cylindrical silos with corrugated walls without stiffeners during filling. The behaviour of bulk solid was described with a hypoplastic constitutive model. Non-linear FE analyses with both geometric and material non-linearity were performed. The numerical results were compared with the Eurocode formulae. The quantitative estimation of the strengthening effect of the bulk solid stiffness on the silo stability, buckling strength and wall stresses was performed.

1   INTRODUCTION

Thin-walled corrugated cylindrical metal silos are frequently used in the industry to store different bulk solids. They are usually strengthened by vertical stiffeners distributed uniformly around the silo circumference. Without vertical columns they are rarely used due to the lack of vertical stiffeners preventing wall buckling and their high sensitivity to wind action. These silos are however more economical with respect to the steel consumption which can make them commercially attractive. There do not exist reliable formulae for dimensioning these silo types. Eurocode 3 includes simple formulae to calculate the buckling strength which provide a very conservative approach since do not consider a positive bedding effect of solids. It is well known that the buckling strength of silo shells containing bulk solids at rest may significantly be enhanced when compared to empty silos due to both internal pressure and lateral support produced by the silo fill. On one hand, the internal pressure in bulk solids acting on the wall straightens wall imperfections and increases the buckling strength. On the other hand, the silo walls are supported by the fill which restraints them against buckling.

The behaviour of two cylindrical metal silos of a different size with corrugated sheets and without stiffeners containing sand was investigated during silo filling [Kuczynska et al. 2015]. The results were compared with the analytical solutions according to Eurocode 1 and 3. Three-dimensional linear and non-linear stability Finite Element (FE) analyses were carried out without initial geometric imperfections. In order to describe the bulk solid behaviour, a hypoplastic constitutive model was used.

2   HYPOPLASTIC CONSTITUTIVE MODEL

The FE-analyses were carried out with a hypoplastic constitutive model [von Wolffersdorf 1996]. This constitutive model describes the evolution of the effective stress tensor with the evolution of the deformation rate tensor by isotropic linear and non-linear tensorial functions. The hypoplastic model is capable of describing some salient properties of granular materials, e.g. non-linear stress-strain relationship, dilatant and contractant volumetric change, stress level dependence, density dependence and strain softening. In contrast to some conventional elasto-plastic models, a decomposition of deformation into elastic and plastic parts, the formulation of a yield surface, plastic potential, flow rule and hardening rule are not needed. The hallmarks of this model are its simple formulation and procedure for determining material parameters with standard laboratory experiments. A modified hypoplastic relationship proposed by Niemunis and Herle [Niemunis et al. 1997] was used to avoid ratchetting deformation and to improve the small strain behaviour. Due to dynamic stability analyses, we used the hypoplastic constitutive model enhanced by the so-called intergranular strain concept which enabled to model elastic effects in granular materials.

The silo FE simulations were carried out with the following material constants for so-called 'Karlsruhe' sand: $e_{i0} = 1.0$, $e_{d0} = 0.55$, $e_{c0} = 0.84$, $\boxtimes_c = 30°$, $h_s = 5800$ MPa, $\beta = 1$, $n = 0.28$, $\alpha = 0.13$, $m_R = 9$, $m_T = 0.5$, $R = 5 \times 10^{-5}$, $\beta_r = 0.3$ and $\chi = 1$. The solid was assumed to be initially medium dense (initial void ratio $e_o = 0.60$).

3 INPUT DATA IN FE ANALYSES

The 3D silo FE calculations were carried out using a commercial finite element package ABAQUS [Hibbit et al. 2010]. Two different silo sizes were investigated. The medium-size silo (Fig. 1) was $H = 8.4$ m high, its diameter was $D = 2.67$ m and wall thickness $t = 1$ mm. The height of the small-size silo was arbitrarily chosen: height $H = 2$ m, diameter $D = 1$ m and wall thickness $t = 0.1$ and 1 mm. The 4-node thin shell elements S4R with a reduced integration point were employed to represent the silos walls and 8-node linear brick elements C3D8 to depict the bulk solid. In the medium-size silo, the size of the shell elements was 1.0×15 cm^3 and the size of the solid elements was $1.0 \times 15 \times 15$ cm^3. In the small-size silo, the size of the S4R elements was 0.8×5.0 cm^3 and the size of the C3D8 elements was $0.8 \times 5.0 \times 5.0$ cm^3. The total number of the finite elements was 287'000 (45'000 – silo wall and 242'000 – bulk solid) for the medium-size silo and 137'000 (18'000 – silo cylinder and 119'000 - bulk solid) for the small-size silo. The steel was assumed to be elastic or elastic-perfectly plastic. Between the corrugated wall and bulk solid, the Coulomb friction was assumed with the friction coefficient $\mu = 0.3$.

The two types of analyses were performed: stress linear analyses (called here SLA) and geometrically and materially non-linear stability analyses by tracing a force-deflection path (called here GMNA) using the implicit dynamic approach.

In the FE calculations, the silo was loaded using two different methods. In the first case, the empty silo was loaded according to the requirements of Eurocode 1. The maximum load factor λ was defined as the ratio between the maximum calculated wall pressures and wall pressures by Eurocode 1. In the second case, the silo was loaded by the bulk solid described by a hypoplastic constitutive model. The maximum load factor λ was defined as the ratio between the maximum calculated and real volumetric weight of sand Two different load fac-

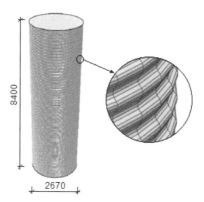

Figure 1.   The dimensions and FE mesh of medium-size silo containing bulk solid assumed for FE calculations (dimensions are in [mm]).

tors were determined: λ_{pl} – the load factor when the silo wall was plasticized and $\lambda_{b,max}$ – the maximum stability load factor.

4 CALCULATIONS OF WALL PRESSURES IN MEDIUM-SIZE SILO

4.1 Empty silo with solid loads by Eurocode 1

The normal p_{hf} and tangential p_{wf} wall pressure during silo filling was calculated according to Eurocode 1 (following the so-called slice-method by Janssen). The maximum values were: $p_{hf} = 21.2$ kPa and $p_{wf} = 10.6$ kPa for medium-dense sand ($e_o = 0.60$, $\gamma = 16.75$ kN/m³). The wall pressures were calculated for the standard wall friction coefficient $\mu = 0.6$ (for the corrugated wall). The maximum horizontal tensile stress in the corrugated wall was 26 MPa and did not exceeded the steel yield strength equal to 355 MPa.

The wall buckling force in cylindrical silos made from corrugated sheets without stiffeners caused by axial compression was calculated according to Eurocode 3 and it was equal to $n_{x,Rk} = 9.86$ kN/m. The standard buckling load factor of the silo during filling was only $\lambda_b = 0.14$. Thus the permissible buckling force of the corrugated wall for the lowest ring was exceeded 7 times (maximum vertical force $n_x = 73.0$ kN/m). In order to satisfy the standard buckling conditions, the wall has to be 3 mm thick (instead of $t = 1$ mm).

Figure 2 presents the FE results of stresses (SLA) in the silo loaded by sand following Eurocode 1. With respect to the wall corrugation, the calculated distribution of the circumferential wall stresses in the silo wall was significantly different than this given by Eurocode 3 formulae which assumed solely the tensile stresses in a circumferential direction. The calculated wall stresses were both compressive and tensile. The stress was higher close to the sheet wave peak and lower close to the sheet wave valley (Fig. 2A). The calculated maximum vertical forces in the corrugated wall were $R_F = 73.2$ kN/m and were consistent with the analytical outcomes. The yield strength ($f_y = 355$ MPa) was exceeded maximum 4.7 and 10 times in the circumferential and meridional direction; the maximum circumferential stress was 1600 MPa and the meridional one was 3350 MPa) (Fig. 2A).

4.2 FE calculations for silo containing sand

The distributions of the wall normal pressure p_{hf} and wall wall traction p_{wf} from FE calculations of the silo containing solid (Fig. 3) differ noticeably from the standard solutions.

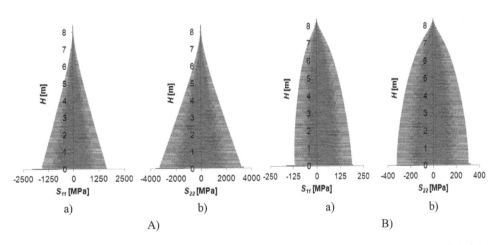

Figure 2. Distribution of circumferential S_{11} (a) and meridional S_{22} (b) wall stresses along silo height from FE calculations (SLA) in medium-size silo of Figure 1 with corrugated walls loaded by Eurocode 1 (A) and with sand described with hypoplastic constitutive model (B) ('+' - tension, '-' – compression).

The calculated maximum wall pressure in the lower part of the silo was higher by 95% for sand than by the standard. Due to the fact that the resultant vertical traction (and thus the resultant vertical force) had a small value, the normal wall pressure distribution was almost hydrostatic (Fig. 3a). The vertical wall traction had also a completely different distribution (Fig. 3b) as compared with Eurocode 1 that resulted in a significant reduction of the vertical reaction force ($R_F = 54.9$ kN (6.55 kN/m)), since the significant part of the vertical force was transferred to the silo bottom. The maximum wall stresses occurred in the silo bottom part: $S_{11} = 195$ MPa and $S_{22} = 325$ MPa. They were 8–10 times lower than the wall stresses for the silo loaded by Eurocode 1 (Fig. 2B).

5   STABILITY CALCULATIONS OF MEDIUM AND SMALL-SIZE SILO

The stability calculation results evidently show that the silo is stable ($\lambda_b > 1$) if the solid is described by the hypoplastic constitutive model. Despite the similar vertical reaction force (Fig. 4), the strength of the silo with the bulk solid was 22 times higher for sand ($\lambda_{b,max} = 3.3$)

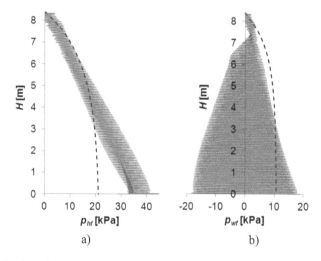

a)   b)

Figure 3.   Distribution of normal wall pressure p_{hf} (a) and vertical wall traction p_{wf} (b) along height from FE calculations (SLA) in medium-size silo for sand (described by hypoplastic constitutive model) (dashed line—Eurocode 1 and continuous line—FE results).

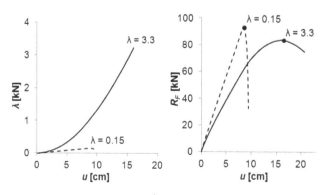

Figure 4.   Evolution of buckling load factor λ and vertical reaction wall force R_F versus vertical displacement u of top of silo with sand (dashed line—Eurocode 1 and continuous line—FE result).

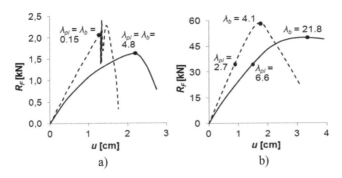

Figure 5.   The vertical wall reaction force R_F versus vertical displacement of silo top edge u in small-size silo with sand for different wall thickness t: empty silo loaded according to Eurocode 1—dashed line and silo loaded by real solid—continuous line (a) $t = 0.1$ mm, b) $t = 1$ mm).

as compared with the analytical formulae in Eurocode 3 ($\lambda_{b,max} = 0.15$). For the both loading ways, the yield strength was exceeded first at the peak of corrugated wall waves and the wall started to deform as a 'harmonica'. For the silo loaded according to Eurocode 1, deformation occurred mainly in the lower part of the silo that resulted in the faster wall stability loss (at $\lambda_{b,max} = 0.15$). For the silo loaded by a real bulk solid, the silo deformation occurred more uniformly over the entire wall height than in the empty silo that resulted in the higher silo strength ($\lambda_{b,max} = 3.3$).

Figure 5 presents the evolution of the wall reaction force R_F versus the vertical displacement of the silo top edge u and calculated load factors λ for the different wall thickness t in the small-size silo loaded according to Eurocode 1 and loaded by real sand described with hypoplastic constitutive model (Fig. 5). The significant strengthening effect was observed for the silo with sand for both corrugated sheet thicknesses t. The strengthening stability effect caused by the presence of sand was higher for $t = 0.1$ mm (3000%) than for $t = 1.0$ mm (400%). The silo containing sand exhibited also a higher and more stable post-buckling behavior without fluctuations as compared to the empty silo loaded according to Eurocode 1.

6   CONCLUSIONS

The obtained results of a small and medium-size silo without stiffeners containing sand show that sand significantly increases the buckling strength of silos during filling as compared to the silos loaded by Eurocode 1. The strengthening effect was related to the wall thickness, it was greater when the silo wall was thinner. For the medium-size silo, the strengthening effect of the solid was 2000% ($t = 1$ mm). When the small-size silo was loaded by sand, this effect was 3000% with the wall thickness $t = 0.1$ mm and 400% with $t = 1$ mm.

The distribution of the normal wall pressure and vertical wall traction significantly differed from Eurocode 1. The distribution of the wall normal pressure was almost hydrostatic. The maximum wall pressure at the bottom was higher even by about 50% as compared to Eurocode 1. The vertical wall traction was lower as compared with the Eurocode that caused a significant reduction of the vertical reaction force (1000%-decrease in the medium-size silo).

ACKNOWLEDGMENT

We acknowledge the support from Grant WND-POIG.01.03.01–00–099/12–01 financed by Polish National Centre for Research and Development NCBiR (2013–2015).

REFERENCES

Eurocode 1, BS EN 1991–4: Actions on Structures. Part 4: Silos and Tanks. General Principles and Actions for the Structural Design of Tanks and Silos; 2009.

Eurocode 3, BS EN 1993–4-1: Design of Steel Structures. Part 4–1: Silos, Tanks and Pipelines-Silos; 2007.

Hibbit, Karlsson, Sorensen Inc. *ABAQUS. Theory Manual, version 6.10*, 2010.

Kuczyńska, N., Wójcik, M., Tejchman, J. Effect of bulk solid on strength of cylindrical corrugated silos during filling. *Journal of Constructional Steel Research* 2015, 115, 1–17.

Niemunis A, Herle I. Hypoplastic model for cohesionless soils with elastic strain range. *Mechanics of Cohesive-Frictional Materials* 1997:2:279–299.

von Wolffersdorf PA. A hypoplastic relation for granular materials with the predefined limit state surface. *Mechanics of Cohesive-Frictional Materials* 1996;1:251–271.

Hydraulic Engineering IV – Xie (Ed.)
© 2016 Taylor & Francis Group, London, ISBN 978-1-138-02948-4

Trend changes in resistivity image: A new indicator of earthquakes

T. Zhu, B. Zhang & J. Zhou
Key Laboratory of Seismic Observation and Geophysical Imaging, Institute of Geophysics, China Earthquake Administration, Beijing, China

ABSTRACT: We first acquire the resistivity of two sets of man-made samples using electrical resistivity tomography under different stresses and then construct their images of relative change. It was found that the resistivity-increased region expanded; meanwhile, the resistivity-decreased region shrank step by step with the increase of uniaxial stress in all of the resistivity images corresponding to the measuring lines placed at different directions (parallel to, perpendicular to, and intersecting the axis of load at 45°). These features are very similar to those prior to Tangshan ML 4.4 and 5.0 earthquakes that occurred on April 14, 1998, i.e. the similar trend changes in resistivity image that occurred during the preparation of the two earthquakes. This new finding encourages us to conclude that trend changes in resistivity image will be a potential, effective, and reliable indicator of an earthquake.

1 INTRODUCTION

Seismic activities are very good for the planet we live on, but disastrous for human beings, especially catastrophic earthquakes such as the 2008 Wenchuan Ms 8.0 earthquake (Dai *et al.* 2011), the 2004 Indonesia Mw 9.0–9.3 earthquake (Løvholt *et al.* 2012), and the Japan Mw 9.0 earthquake (Ozawa *et al.* 2011). One way to reduce the loss and harm due to earthquakes as far as possible is to find the anomalies associated with an earthquake and then to predict future earthquakes. Earth resistivity anomaly has been verified as a sort of precursor of an earthquake (Lu *et al.* 1999). However, to date, only a fixed Schlumberger sounding array consisting of a pair of current and of potential ones in general has been used for detecting the direct current Earth resistivity anomaly associated with an earthquake all over the world, which can only record general resistivity anomalies from a certain depth upwardly to the Earth's surface with time. This is because the resistivity anomaly of the uppermost subsurface (here refers to the range from the Earth's surface downwardly to about 20 m depth) is easily affected by human activities (e.g. irrigation and pumping) and climate (e.g. precipitation and drought), and thus it is very difficult to be separated from the general anomaly. In addition, due to the heterogeneity of the electrical structure, the resistivity at a certain depth will probably increase or decrease due to stress, which probably results in the general resistivity that will not change greatly during the preparation of an earthquake, i.e. it is not possible to detect the Earth resistivity precursor of an earthquake, even for a catastrophic earthquake such as the 2008 Wenchuan Ms 8.0 earthquake (Zhang *et al.* 2009). Many investigators have tested new observation techniques such as deep-well (Kang *et al.* 2013) and multi-separation array observation (Wang *et al.* 2011), in order to attempt to address this issue, but their effectiveness remains unclear. Here, based on electrical resistivity tomography (ERT) from the geophysical survey, we conducted the experiments on resistivity changes associated with stress in the laboratory. The number of resistivity acquired by ERT allows us to construct a resistivity image, which gives us a chance to study the changes in resistivity image, but not as a single value (Wang *et al.* 1975), associated with stress.

2   EXPERIMENTAL PROCEDURE

Two sets of man-made samples, uniform (MUR) and high (MHR) resistivity ones, were used in the present study. Each set of the samples contained two cubic specimens with a side length of 30 cm. Each specimen was subjected to the following procedure. First, we uniformly mixed the ordinary Portland cement 42.5 R, river sand, white Portland cement 42.5, and salt at the weight ratio of 11:11:2.2:1 and poured into the mold of the specimen. Second, four common midpoint measuring lines L1–L4 with the electrode spacing of 5 mm were arranged on one free surface of the specimen (Figure 1a). The intersection angle of two adjacent measuring lines was 45°. Finally, the specimen was placed in the air at least for 1 month to make it consolidated enough. The specimen had the compression strength of about 44.4 MPa and density of about 1900 kg/m³. The MUR sample was made only using the above-mentioned uniform mixture. Each MHR sample was embedded with a high resistivity block, which was a cuboid of the pure ordinary Portland cement 42.5 with the dimension of $5 \times 2 \times 2$ cm placed at 2.5 cm just beneath L1 (Figure 1a).

The experimental system is shown in Figure 1b. The uniaxial servo control YAW-5000F loading machine (http://shijin.shuoyi.com/shtml/shijin/product/57fd78bf28f0244d.html) manufactured by Jinan Shijin Group Co. Ltd was used for the axial load. The apparent resistivity data were acquired through a DC resistivity meter, which was specially designed for measuring the resistivity of the rock sample in the laboratory. Its technological specifications are the same as those of DCX-1 DC resistivity meter (http://www.kuyibu.com/gongy-ing/1413519.html), except for its input impedance of about 4104 MΩ, and the accuracy of the potential measurement of about 5% for the samples used in this study. The Wenner-α array was used for data acquisition.

The samples were compressed by the uniaxial load (Figure 1b). First, the stress was increased step by step to a certain level at a uniform rate, and resistivity measurements were performed along the measuring lines respectively at each stress. For instance, we conducted resistivity measurements at increasing stress values of 0, 11.11, 22.22, 27.77, and 33.33 MPa at the uniform rate of 2.0×10^{-2} MPa/s for the MUR sample and of 0, 11.11, 22.22, 33.33, and 40 MPa at the uniform rate of 2.0×10^{-2} MPa/s for the MHR sample. The parameters of the measuring lines are listed in Table 1.

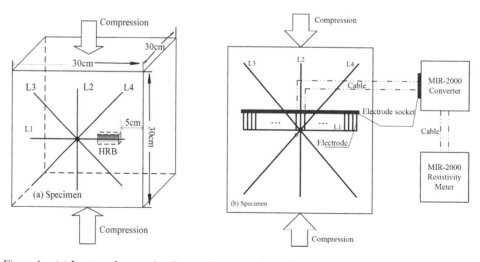

Figure 1.   (a) Layout of measuring lines and location of the high resistivity block (HRB), and (b) the experimental system.

Table 1. Parameters of measuring lines.

ML	NE	SE	NF	Remarks
L1	43	5 mm	14	Perpendicular to the axis of load
L2	43	5 mm	14	Parallel to the axis of load
L3	57	5 mm	18	Intersecting northeastwardly the axis of load at 45°
L4	57	5 mm	18	Intersecting northwestwardly the axis of load at 45°

Note: ML, NE, SE, and NF indicate the measuring line, the number of electrodes, spacing of the electrode, and "n" factors (Diaferia et al., 2006), respectively.

3 RESULTS AND ANALYSIS

Here, we study the relative change $\Delta\rho_{NR}$ of apparent resistivity, which can be calculated using the following equation:

$$\Delta\rho_{NR} = (\rho_N - \rho_0) \times 100\% / \rho_N,\tag{1}$$

where ρ_N and ρ_0 represent the measured apparent resistivity values at the stress value of N MPa and 0 (zero) MPa.

As can be seen from Figure 2, the RRCIs could be divided simply into the Resistivity-Decreased Region (RDR) (blue to white) and the Resistivity-Increased Region (RIR) (white to red). When the stress is less than 33.33 MPa for L1 of the MUR sample (Figure 2a), with the increase in stress, the RDR shrinks while the RIR expands gradually. The average resistivity value of the RDR almost remains constant while that of the RIR increases gradually. When the stress is less than 22.22 MPa for L2–L4 of the MUR sample (Figure 2b, c, and d), RDRs occupy most of RRCIs. Meanwhile, with the increase in stress to 33.33 MPa, RDRs shrink while RIRs expand rapidly, which results in the rapid increase in the average resistivity values of the RIR.

In Figure 2b, c, and d, the RIR and the RDR in RRCIs of the MHR sample (not presented because of the similarity) have the same change in behaviors associated with stress as those of the MUR sample. However, the embedded high-resistivity block has a great influence on the increased resistivity amplitude just in its location and surroundings, especially when the stress is more than 22.22 MPa, indicating that the presence of a high-resistivity block may make the changes in resistivity associated with stress more prominent.

4 DISCUSSION AND CONCLUSION

The RRCIs along the four directions of each specimen have the same trend change associated with stress (Figure 2): the RDRs (blue to white) will shrink and RIRs (white to walnut) will expand step by step with the increase in stress. These phenomena are very similar to the results observed near Changli station, Hebei Province, China, prior to Tangshan ML 4.4 and 5.0 earthquakes that occurred on April 14, 1998 (Feng et al. 2001): the RDR (RIR) had been expanding (shrank) step by step since 9 months, and then, the phenomena disappeared about 2 months after the earthquake. In addition, it was found that climate might have affected the magnitude of Earth resistivity; however, it has a little influence on the trend change in resistivity images (Figure 3). These results indicate and encourage us to conclude that the intuitive and regular characteristics of changes in resistivity image would result in identifying the Earth resistivity anomaly related to an earthquake easily and more reliably. This advantage will probably impel it to be a simple (easy to identify whether it is a precursor of an earthquake or not) and potential effective (expansion or shrinkage of RDRs or RIRs always associated with an earthquake) indicator of an earthquake.

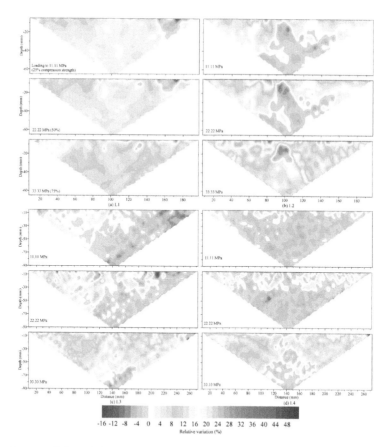

Figure 2.   Images of the relative resistivity change corresponding to the measuring lines at different stress values for the MUR sample.

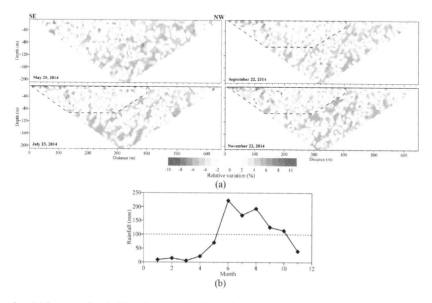

Figure 3.   (a) Images of resistivity changes relative to those on March 20, 2014, at the site on Xiaojiang fault near Xiaoxinjie, Yunnan province, China, and (b) the curve of average monthly rainfall recorded at Songming station very close to our test site.

As can be seen in Figure 3b, the rainy season near our site began in May and ended in October in 2014, and the rainfall was more than 100 mm (marked by dotted line) from June to October. Our resistivity images (Figure 3a) revealed the change in rainfall patterns. For example, the low resistivity areas (enclosed by dash line) that were shallower than about 100 m in the images on July 23, September 22, and November 22 were obviously due to rainfall. However, the changes in resistivity images (Figure 3a) distributed randomly, i.e. there was no trend change (RDR or RIR expanded or shrank step by step), which indicated that rainfall may affect the magnitude of resistivity, but has a little influence on the trend change in resistivity image.

ACKNOWLEDGMENT

This work was supported by the National Natural Science Foundation of China (NSFC, Grant 41574083) and the Special Project for the Fundamental R&D of the Institute (DQJB13B05).

REFERENCES

Dai, F.C., Xu C., Yao X., Xu L., Tu, X.B. & Gong, Q.M. 2011. Spatial distribution of landslides triggered by the 2008 Ms 8.0 Wenchuan earthquake, China. *Journal of Asian Earth Sciences* 40: 883–895.

Feng, R., Hao, J.Q. & Zhou, J.G. 2001. Resistivity tomography for earthquake monitoring. *Chinese Journal of Geophysics* 44(6): 819–830.

Kang, Y., An, H., Ma, K. & Tan, D. 2013. Test analysis on geoelectrical resistivity observation combining the surface and deep-well methods at Tianshui seismic station in Gansu province. *China Earthquake Engineering Journal* 35(1): 190–195.

Løvholt, F., Kühn, D., Bungum, H., Harbitz, C.B. & Glimsdal, S. 2012. Historical tsunamis and present tsunami hazard in eastern Indonesia and the southern Philippines. *Journal of Geophysical Research* 117(B09310): 1–19.

Lu, J., Qian, F.Y. & Zhao, Y.L. 1999. Sensitivity analysis of the Schlumberger monitoring array: application to changes of resistivity prior to the 1976 earthquake in Tangshan, China. *Tectonophysics* 307: 397–405

Ozawa, S., Nishimura T., Suito, H., Kobayashi, T., Tobita, M. & Imakiire T. 2011. Coseismic and post-seismic slip of the 2011 magnitude-9 Tohoku-Oki earthquake. *Nature* 475: 373–376.

Wang, C.Y., Goodman, R.E., Sundaram, P.N. & Morrison, H.F. 1975. Electrical resistivity of granite in frictional sliding: Application to earthquake prediction. *Geophysical Research Letters* 2: 525–528.

Wang, L., Zhu X., Zhu, T., Zhang, S., Liu, D., Hu, Z. & Zhang, Y. 2011. Multi-separation array geoelectrical resistivity observation system and its experimental observation. *Earthquake* 31: 20–31.

Zhang, X.M., Li, M. & Guan, H.P. 2009. Anomaly analysis of earth resistivity observations before the wenchuan earthquake. *Earthquake* 29: 108–115.

Hydraulic Engineering IV – Xie (Ed.)
© 2016 Taylor & Francis Group, London, ISBN 978-1-138-02948-4

Study on the application of blending Ground Sand Dust with unqualified fly ash in RCC

Guo-xin Chen, Ye-ran Zhu & Guo-hong Huang
Nanjing Hydraulic Research Institute, Nanjing, China

ABSTRACT: The content of f-CaO in high calcium Fly Ash (FA) produced by Kaiyuan Xiaolongtan Power Station reached at 9.32% and that of SO_3 was higher than the state criterion. The content of f-CaO and SO_3 can be effectively reduced by blending Ground Sand Dust (GSD) with FA, so the illegibility of cement soundness can be avoided. The study showed that: 1) ground in trial ball mill for 60 min, the river sand can be made into GSD with fineness (45 μm) of 12.8% and the specific surface area of 12,656.44 cm²/g; 2) while GSD was blended with FA, the 28d compressive strength declines with the dosage of GSD was added; 3) cement soundness was up to state criterion when the blending rate of GSD ranges from 20 to 50% in the mineral admixture; 4) GSD ground for 60 min had the best performance in $C_{180}20$ RCC; 5) as to the different GSD blending rates, 30% was the best with only a 5.4% loss of 90d compressive strength; and 6) due to the lower proportion of SD in river sand, the workability and density of concrete is poor. The performance of the concrete can be evidently improved if 10% sand was replaced by GSD. The 90d compressive strength can also be increased by 35.5%.

1 INTRODUCTION

With the rapid development of Western China, more and more hydraulic projects are constructed at the same time. It directly leads to the lack of building materials, especially to the lack of qualified FA. It was the same situation in Yunnan province. Due to the higher traffic expense, the supply of FA became tenser and the prices became higher. While those FA produced by Xiaolongtan Power Station have the features of high CaO content especially high f-CaO content and high SO_3 content, it belongs to unqualified FA [WU Xue-li, 2001; FENG Xiao-dong, 1999; LIU Lun-jun, 2010; HUANG Guo-hong, 2006]. According to the current State Criterion, it cannot be directly applied in the hydraulic projects. Too many wasted FA also will be idle of natural resources and a severe environmental problem. If the GSD is used as a new type of mineral admixture to replace part of FA, the excessive f-CaO and SO_3 content will be devalued. So the unqualified FA may be used not only to cut down the cost of construction, but also to reduce environmental pollution.

In this paper, the high f-CaO and high SO_3 content FA will be replaced by different fineness GSD in different dosages; then the soundness of blending mineral admixture and its performances in RCC will be discussed to investigate the feasibility of its practical appliance.

2 MATERIALS AND METHODS

2.1 *Experimental materials*

1. Cement: Ziyan P.O.32.5 cement, produced by Yunnan Ziyan Cement Co., Ltd., its property was presented in Table 1–Table 2.

2. FA: Produced by Yunnan Xiaolongtan Power Station. Its density, fineness, water demand ratio, and moisture content were 2.84 g/cm³, 22.2%, 93%, and 0.1%, respectively. Its chemical composition was shown in Table 3. It could be concluded from the above data that this kind of FA belonged to high calcium FA, its CaO and f-CaO content was much higher than ordinary FA; its water demand ratio and ignition loss was fitted for grade FA, its fineness could only be fit for grade FA, while the SO_3 content exceeded the upper limit of 3% in the State Criterion. Generally speaking, it belonged to unqualified FA.

3. Fine aggregate: Natural river sand from Honghe River, its apparent density, saturated surface dry moisture content, fineness moduli, and stone dust content was 2,670 kg/m³, 1.36%, 2.9, and 2.1%, respectively.

4. Coarse aggregate: Pebble of Honghe River. Big pebble (40 mm~80 mm), apparent density is 2,730 kg/m³, moisture content is 0.90%; middle pebble (20 mm~40 mm), apparent density is 2,770 kg/m³ and moisture content is 1.08%; small pebble (5 mm~20 mm), apparent density is 2,740 kg/m³ and moisture content is 0.77%. The ratio of big, middle, and small pebble in three gradations was 3:4:3. The middle and small pebble in two gradations was 5:5.

5. GSD: It was ground with natural river sand in trial ball mill; the chemical composition of GSD was shown in Table 4. It was found that the major composition of GSD was SiO_2 and the SO_3 content was far lower than FA, the SO_3 content of mineral admixture could be effectively reduced by replacing FA with different dosages of GSD, and which was definitely smaller than 3% when the dosage of GSD was bigger than 25%. Although, because of the low content of active composition, the early age strength of concrete would slightly decrease.

6. Admixture: HLC-NAF retardant pumping agent (RCC type, liquid, solid content 30%) and HK-A air-entraining admixture (liquid, solid content 2.5%); all produced by Nanjing Ruidi High-tech Co., Ltd.

Table 1. Chemical composition of cement (%).

CaO	SiO_2	Al_2O_3	Na_2O	K_2O	MgO	SO_3	Fe_2O_3	Loss
59.90	18.82	6.66	0.07	0.35	1.38	2.91	3.82	3.47

Table 2. Physical and mechanical property of cement.

Density (g/cm³)	Fineness (%)	Consistency (%)	Setting time (hr: min)		Compressive strength (MPa)		Flexural strength (MPa)	
			initial	final	3d	28d	3d	28d
3.06	4.40	24.0	3:16	4:30	22.8	46.5	5.9	11.6

Table 3. Chemical composition of FA (%).

CaO	SiO_2	Al_2O_3	Na_2O	K_2O	MgO	SO_3	Fe_2O_3	f -CaO	Loss
34.00	28.38	13.94	0.07	0.56	2.92	3.83	10.88	9.32	2.11

Table 4. Chemical composition of GSD (%).

Loss	CaO	SiO_2	Al_2O_3	Na_2O	K_2O	MgO	SO_3	Fe_2O_3
5.96	6.61	71.94	6.55	0.75	1.20	2.06	0.14	2.69

2.2 Experimental methods

2.2.1 Preparation and fineness test of GSD

SYM¢500 × 500 trial ball mill was used to prepare GSD; the fineness of samples was tested at a predetermined grinding time. The weight of screen residue and specific surface area were tested to confirm the fineness of GSD, which used reverse-pressure screen and laser particle size analyzer, respectively.

2.2.2 Soundness test

The soundness test was according to the soundness test method (standard method) in *GB/T 1346–2001*. This is the method to judge the soundness based on the Le Chatelier's expansion value.

3 RESULTS AND DISCUSSION

3.1 Influence of different grinding time to the fineness and properties of GSD

The relationship of fineness and milling time of GSD was presented in Table 5. It was shown that the fineness of GSD turned out abnormal at 90 min of grinding. It can be speculated due to the absence of grinding aids in the process of grinding, the specific surface energy of particles augmented with the prolonging of grinding time, so the primary particles agglomerated again. It was also found that the average specific surface area of GSD particles after 15 min~90 min of grinding time was 7,566 cm²/g, 8,887 cm²/g, 12,656 cm²/g, and 11,830 cm²/g, respectively. As the result of the weight of the screen residue was revealed, the analysis of the particle size also showed the law that the fineness decreased first and increased after 90 min of grinding time.

The relationship between fluidity and water demand of mortar and the grinding time of GSD was shown in Table 6. It was found that the water demand of GSD is not changeable after 60 min of grinding. So 60 min of grinding will be the suitable time both from technical and economic view.

3.2 Influence of different blending dosage of GSD and FA to the mortar strength

The compressive and flexure strength of mortar with different dosages of GSD and FA were shown in Table 7. It was found that the 28d flexure strength increased a little instead of

Table 5. Fineness of ground sand dust after different time of grinding.

Time of grinding (min)	15	30	60	90
Fineness (45 μm) (%)	–	13.4	12.8	28.6
Fineness (80 μm) (%)	24.0	2.2	2.6	2.2

Table 6. Water demand and mortar fluidity of FA after blending with GSD ground at different time.

No.	C (g)	FA (g)	GSD (g)	Time of grinding (min)	W (g)	Mortar fluidity (mm)
Y1–1	250	–	–	–	134	130
Y1–2	175	75	–	–	130	130
Y1–3	175	–	75	15	135	130
Y1–4	175	–	75	30	137	135
Y1–5	175	–	75	60	138	135
Y1–6	175	–	75	90	138	135

decreasing when the dosage of GSD was lower than 50 wt%. Among different dosages, the flexure strength was highest at 30 wt%. But the 28d compressive strength declined as the dosage of GSD increased. When the dosage of GSD exceeded 50 wt%, both the compressive and flexure strength were decreased apparently.

3.3 Influence of different blending dosages of GSD and FA to the soundness

Cement soundness was tested by measuring the Le Chatelier's expansion value in which different dosages of FA were added. The soundness of cement with different blending rates of SD were also tested when the mineral admixture dosage was fixed at 60 wt%. They were all shown in Table 8. It was found that the expansion values were all lower than 3 mm when the dosage of FA was not bigger than 40 wt%, so the soundness were all qualified; which was 4 mm when the dosage of FA was 50 wt%, the soundness may be unqualified with the fluctuation of FA quality; and it was bigger than 5 mm when the dosage reached 60 wt%, so the soundness was unqualified. It was also found that when the mineral admixture was fixed at 60 wt%, the soundness was qualified with the blending rate of GSD in the range of 20~50 wt%.

3.4 Influence to the RCC performances after blending GSD with FA

In Table 9, the GSD after different time of grinding was blended with FA in dosage of 30 wt% in some $C_{180}15$ RCC mix proportion. It was found that when GSD after 60 min of grinding

Table 7. Mortar strength of cement after blending FA with GSD.

No.	C (g)	FA (g)	GSD (g)	Ratio of GSD (%)	28d compressive strength (MPa)	28d flexural strength (MPa)
Y2–1	450	–	–	–	46.6/100	9.8/100
Y2–2	315	135	–	0	37.2/80	8.7/89
Y2–3	315	108	27	20	35.1/75	9.0/92
Y2–4	315	94.5	40.5	30	33.2/71	9.4/96
Y2–5	315	81	54	40	31.4/67	9.0/92
Y2–6	315	67.5	67.5	50	30.7/66	8.9/91
Y2–7	315	27	108	80	26.6/57	7.9/81
Y2–8	315	–	135	100	24.9/53	7.5/76

Table 8. Soundness of cement with different dosage of FA and GSD.

No.	Composition of cement paste (g) C	FA	GSD	W	Dosage of mineral admixtue (%)	Ratio of GSD (%)	Initial clearance (mm) 1	2	Clearance after boiling (mm) 1	2	Average accretion (mm)
Y3–1	500	0	0	118	0	0	10.5	10.0	11.0	12.5	1.5
Y3–2	450	50	0	117	10	0	10.0	11.0	11.5	12.0	1.3
Y3–3	400	100	0	116	20	0	11.0	11.5	12.0	14.0	1.2
Y3–4	350	150	0	116	30	0	9.0	10.0	11.5	12.5	2.5
Y3–5	300	200	0	115	40	0	10.0	10.5	12.0	13.0	2.3
Y3–6	250	250	0	114	50	0	10.5	14.5	9.5	13.5	4.0
Y3–7	200	300	0	113	60	0	10.0	13.0	17.0	21.0	7.5
Y3–8	200	240	60	114	60	20	10.0	11.0	11.5	13.5	2.0
Y3–9	200	210	90	114	60	30	11.5	13.0	13.5	14.0	1.5
Y3–10	200	180	120	115	60	40	11.0	11.5	12.5	14.5	2.3
Y3–11	200	150	150	116	60	50	9.5	9.5	10.5	10.0	0.8

Table 9. Performances of GSD with different grinding time in $C_{180}15$ RCC.

No.	Mix proportion (kg/m³)						Time of grinding (min)	VC Value (sec)	Entrained air content (%)	Volume weight (kg/m³)	Compressive strength (MPa)	
	W	C	FA	GSD	S	G					7d	28d
Y4–1	62	56	84	–	744	1446	–	11.5	2.8	2380	6.3	12.8
Y4–2	62	56	58	26	744	1446	15	9.7	3.0	2364	6.1	11.7
Y4–3	62	56	58	26	744	1446	30	8.2	3.2	2351	7.5	13.3
Y4–4	62	56	58	26	744	1446	60	7.8	3.4	2351	7.6	13.5
Y4–5	62	56	58	26	744	1446	90	4.7	3.8	2337	6.7	12.3

Table 10. Mix proportions and performances of $C_{180}20$ RCC.

No.	Mix proportion (kg/m³)						Water reducer (%)	VC Value (sec)	Entrained air content (%)	Compressive strength (MPa)		
	W	C	FA	GSD	S	G				28d	90d	180d
Y5–1	76	84	68	–	711	1510	1.4	5.8	2.0	16.8	23.7	25.9
Y5–2		84	54	14	711	1510	1.4	6.4	2.5	13.1	18.9	23.7
Y5–3		84	48	20	711	1510	1.6	8.6	2.6	15.6	21.0	24.5
Y5–4		84	41	27	711	1510	1.7	5.0	2.8	13.3	20.7	22.1
Y5–5		84	34	34	711	1510	1.8	7.0	2.5	11.6	15.1	18.2
Y5–6		84	48	91	640	1510	1.9	10.6	2.3	20.8	29.9	33.2

time was used, the VC value of the concrete mixture was the least, the compressive strength was the most, and the entrained air content and volume weight was moderate. So the integrated effect was the best of 60 min GSD.

In Table 10, the GSD after 60 min of grinding time was chosen to blend with FA in the C18020 RCC, the blending rate was 20, 30, 40, and 50 wt%, respectively. The river sand we used had lower SD content, which would lead to the defects such as poor workability and low density, to offset these shortcomings, 10 wt% GSD was used to replace the river sand in the sample Y5–6. It was found that GSD was blended with FA in the dosage of 20 50 wt%, the dosage of HLC-NAF increased accordingly; among different blending rates, the performance of concrete was best at 30 wt%, 90d compressive strength decreased by 5.4% only; the 90d compressive strength decreased quickly by 29.7% at the blending rate of 50 wt%. If 10 wt% GSD was used to replace the river sand, the performance of concrete improved apparently, and 90d compressive strength increased by 35.5%.

4 CONCLUSION

1. Ground in trial ball mill for 60 min, the river sand can be made into GSD with fineness (45μm) of 12.8% and specific surface area of 12,656 cm²/g.
2. When GSD was blended with FA, the 28d compressive strength declines with the dosage of GSD was added accordingly, while 28d flexural strength increases a little instead of decreasing. Among different dosages of GSD, flexural strength was highest at 30 wt%, merely lowered than pure cement paste.
3. Dosage of mineral admixture in cement paste was 60 wt% and blending rate of GSD was 20~50 wt%, soundness of cements were all qualified.
4. As to different grinding time, 60 min was best in C18020 RCC. When GSD was blended with FA in the dosage of 20~50 wt%, the dosage of HLC-NAF increased accordingly; among different blending rates, the performance of concrete was best at 30 wt%, 90d compressive strength decreased by 5.4% only.

5. Due to the low proportion of SD in river sand, the workability and density of concrete is poor. The performance of the concrete can be evidently improved if the 10% sand was replaced by GSD. The 90d compressive strength can also be increased by 35.5%.
6. Blending GSD with unqualified FA at proper scale, the soundness of cement was qualified, the influence of concrete strength was controllable, but its long term volume stability still needs more experiments to validate it.

REFERENCES

Feng Xiao-dong. Study on roller compacted concrete material properties for Feixiangcun Reservoir[J]. Yunnan Water Power, 1999, 3:64–67.

Huang Guo-hong, Lu An-qi, Chen Guo-xin, et al. Effect of fine aggregate on performance of RCC[J]. Water Resources and Hydropower Engineering, 2006, 37(6):54–58.

Liu Lun-jun. Application of ground limestone powder as RCC admixtures[J]. Hunan Water Resources and Hydropower, 2010, 6:56–59.

Wu Xue-li, Yang Qian-rong, GU Zhang-zhao. Characteristics of high calcium fly ash and concrete containing high calcium fly ash[J]. Fly Ash Comprehensive Utilization, 2001, 1:21–25.

Hydraulic Engineering IV – Xie (Ed.)
© 2016 Taylor & Francis Group, London, ISBN 978-1-138-02948-4

The molecular structural design and performance characterization of Concrete Anti-Clay Additives

Long Xiong, Yao Bi, Xing Li, Shaofeng Wang, Jinwen Wang, Juxiang Xing,
Hui Zhou & Minghua Yuan
China Construction Ready Mixed Concrete Co., Ltd., Wuhan, China

ABSTRACT: Through the influence process analysis of clay on the performance of Poly-carboxylate (PCE), a high density cationic polymer has been designed and synthesized. The molecular structure of Concrete Anti-Clay Additives (CACA) was measured by Fourier transform infrared spectrometer. The effect of CACA upon the dispersion of PCE was investigated by mini slump test. The adsorption of PCE and CACA on clays or cement was determined by the method of total organic carbon. The results showed that CACA have obvious resistance to clay due to the strong ability of selective absorption.

1   INTRODUCTION

Polycarboxylate (PCE) was widely used in commodity concrete due to its low dosage, flexible structure design, environmental protection, etc. (Ran, et al. 2009, Plank, et al. 2010, Ota, et al. 1997). Engineering practice has proved, however, that clay in the aggregate can significantly reduce the performance of polycarboxylate, causing the poor liquidity and slump loss of concrete (Liu, et al. 2006, Norvell, et al. 2007, Lei, et al. 2012).

At present, in the process of production when aggregates with excess levels of clay can significantly affect the performance of PCE, the common solutions were washing aggregate, improving the dosage of admixture, control the mud content, etc. Irrigation not only caused environmental pollution and affected gradation, but also not conducive to control the water-cement ratio of concrete. What is more, frozen would bring about construction difficulties in cold season and regions. The improvement of PCE cannot solve the problem of the water reducing rate and slumping loss, sometimes may also bring out the concrete setting time too long.

Therefore, it is of great significance for PCE application to develop a kind of Concrete anti-clay additives widely used in aggregates with high content clay, which can effectively solve the additive dosage volatility, quality instability problem, and the construction difficulties. Based on the interaction mechanism research between admixture and clay mineral, a cationic polymer was designed and synthesized in this paper and the properties of resistance to clay have been characterized by the fluidity test of cement paste and the adsorption test.

2   THE MOLECULAR STRUCTURE DESIGN OF CONCRETE ANTI-CLAY ADDITIVES

Experimental studies have found that only clay with significant water absorption and expansion properties can make an important influence on the concrete performance (Ng, et al. 2012, Atarashi, 2005, Sakai, Atarashi & DAaimon, 2006). The impact process could be described below: 1) At the beginning of the clay hydration, negative charge repulsion causes the expansion between the layers that were produced by dissociation and replaced the cation exchange, then polycarboxylate lose dispersion effect due to its side chain intercalated clay interlayer. 2) Due to the cation gather, free water diffusion into the interlayer of clay. Therefore, the

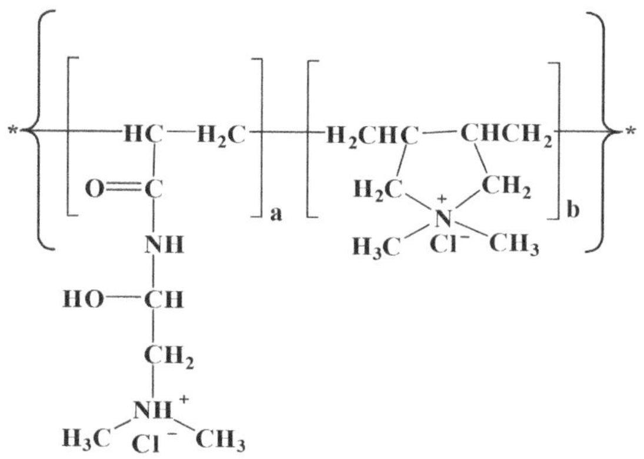

Figure 1. The structure of the CACA samples designed in the study.

failure of polycarboxylate and the decrease of free water in cement pastes lead to the degradation of concrete flow and slump performance (Suter, 2009).

In this paper, high density cationic groups were introduced in the molecular structure of concrete anti-clay additives, which could adsorb quickly on interlayer or the surface of clay by the form of electrostatic attraction, ionic exchange, etc. Consequently, negative charge on clay becomes neutralized. Crystal layer tend to shrink and render it difficult for hydration swelling. Thus, on the one hand, free water accommodated between layers was decreased sharply due to the decrease of the interlayer spacing; on the other hand, target molecules dominated adsorption sites between the layers and produced the steric hindrance that made it difficult for polycarboxylate to adsorb the clay layers. These two aspects could ensure excellent water reducing and slump retention.

To sum it up, the molecular structure of the Concrete Anti-Clay Additives (CACA) design as shown in Figure 1.

3 THE EXPERIMENT

3.1 Synthesis of Concrete Anti-Clay Additives

A certain amount of deionized water and acrylamide were added to a 500 ml four-necked flask equipped with a condenser, a nitrogen inlet, and a stirring bar. It was afterwards stirred for 15 min to give a homogeneous solution. The solution was gently heated to 40°C and the ph value was adjusted to 8~9. Then epichlorohydrin was added to the flask under a fixed feeding rate. After completing the feeding process, the mixture was stirred for half an hour under the same temperature. Then, the pH was adjusted to 4~5. Certain amounts of diallyldimethylammonium chloride and ammonium persulfate were added to the flask and a certain concentration of sodium sulfite solution was supplemented with constant speed. The mixture was incubated at 65°C for a further three hours, and then concrete anti-clay additives were acquired.

3.2 FT-IR

The molecular structure of concrete anti-clay additives was measured by Fourier transform infrared spectrometer (Thermo Nicolet Avatar 370, USA). Set the scan wave number 400–4000 cm^{-1}, scanning times for 50.

3.3 Mini slump test

Test of the fluidity of cement pastes added with PCE was under the national standard GB 8077–2000 "methods for testing uniformity of concrete mixture" and a cement 42.5 R was used supplied by HUAXIN Co., Ltd., which the specific surface area of 3,500 cm²/g, the w/c ratio was 0.29 and the dosage of the PCE was 0.2% (by percent weight based on solids cement). Two kinds of clay were added 4% of the mass of cement with outside mixing way. Concrete anti-clay additives mixed with super plasticizer in a quality percentage of clay.

3.4 Sorption experiments

The adsorption of PCE and concrete anti-clay additives on clays or cement was determined by method of total organic carbon. Details of the test and calculation method could be found in the previous paper (Suter, 2009).

4 RESULTS AND DISCUSSION

4.1 FT-IR

Figure 2 showed the infrared spectrogram of concrete anti-clay additives. As can be seen from the figure, the N – H stretching vibrations were observed in the 3,400 cm⁻¹–3,200 cm⁻¹ region. The stretching vibrations of CH_3 connected to the N⁺ were observed at 2,930 cm⁻¹. The peak of 1,620 cm⁻¹ and 1,660 cm⁻¹ were considered to be caused by C = 0 stretching vibration and N – H deformation vibration, respectively. The stretching vibration and deformation vibration of C–N group in five-membered heterocyclic compound were observed at 678 cm⁻¹ and 1,262 cm⁻¹, respectively. The peak of 980 cm⁻¹ can belong to the characteristic absorption peak of quaternary ammonium groups. The analysis of the FTIR results indicates that the reaction products were a kind of cationic polymers that contain the functional groups of C–N, N–H and five-membered heterocyclic.

4.2 Mini slump test

To ascertain the effect of Concrete Anti-Clay Additives (CACA) on PCE, the fluidity and slump loss behavior of cement slurry containing different dosages of CACA were investigated and shown in Figure 3 and Figure 4, respectively. As can be seen from the Figure 3, the CACA affects the fluidity of cement pastes slightly in the case of without clay. This shows that the CACA have no water reducing effect. While the fluidity of the cement paste increase as the CACA dosage increases in the presence of clay. According to Figure 4, the Slump loss data of cement pastes also show the same trend. This suggests that the CACA have obvious resistance to clay.

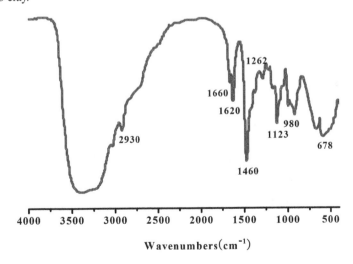

Figure 2.   The infrared spectrogram of Concrete Anti-Clay Additives.

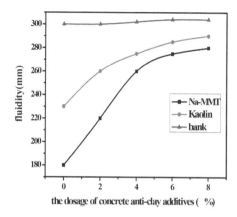

Figure 3. The fluidity of the cement paste with different Concrete Anti-Clay Additives dosage.

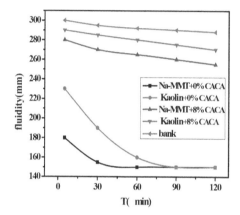

Figure 4. Slump loss behavior of cement slurry with different Concrete Anti-Clay Additives dosage.

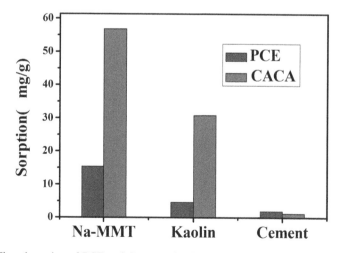

Figure 5. The adsorption of PCE and Concrete Anti-Clay Additives on clays or cement.

4.3 *Adsorption*

As can be seen from the adsorption in Figure 5, clay particles have stronger adsorption capacity of PCE than cement. This is the reason that the PCE lose the dispersion effect. But the clay shows higher adsorption quantity of CACA. This shows that the CACA with cationic structure designed in this article can adsorb the negatively charged surface of clay particles by the way of electrostatic attraction. Therefore, in the system after adding CACA, PCE can be liberated from the clay surface, continue to play a dispersion effect.

5 CONCLUSION

This work confirms that CACA has an obvious effect on clay. The cationic structure designed in this article can adsorb the negatively charged surface of the clay particle by way of electrostatic attraction. On the one hand, free water accommodated between the layers was decreased sharply; on the other hand, target molecules dominated the adsorption sites and produced the steric hindrance, which made it difficult for PCE to adsorb the clay layers. These two aspects could ensure PCE of excellent dispersion performance and slump retention.

REFERENCES

Atarashi D. 2005. Interaction between superplasticizers and clay minerals [J]. *Japan Cement Association*, 58(6): 287–392.
Lei, L. & Plank, J. 2012. A concept for a polycarboxylate superplasticizer possessing enhanced clay tolerance [J]. *Cement and Concrete Research*, (42): 1299–1306.
Liu, X. et al. 2006. Swelling Inhibition by polyglycols in montmorillonite dispersion [J], *Journal of Dispersion Science and Technology*, 25(1):63 66.
Ng, S. & Plank, J. 2012. Interaction mechanisms between Na montmorillonite clay and MPEG-based polycarboxylate superplasticizers [J]. *Cement and Concrete Research*, (42): 847–854.
Norvell, J.K. et al. 2007. Influence of clays and claysized particles on concrete performance [J]. *Journal of Materials in Civil Engineering*, 19(12): 1053–1059.
Ota, A. Sugiyama, T. & Tanaka, Y. 1997. Fluidizing mechanism and application of polycarboxylatete-based superlasticizers [C]. *5th Canmet/Cai Int Conf on superplasticizers and Other Chemical Adimixture in Concrete*: 359–379.
Plank, J. et al. 2010. Fundamental mechanisms for polycarboxylate intercalation into C3 A hydrate phases and the role of sulfate present in cement [J]. *Cement and Concrete Research*, 40(1): 45–57.
Ran, Q. C, et al. 2009. Effect of the length of the side chains of comb-like copolymer dispersants on dispersion and rheological properties of concentrated cement suspensions [J]. *Journal of Colloid and Interface Science*, 336(2): 624–633.
Sakai, E. Atarashi, D & Daimon, M. 2006. Interaction between superplasticizers and clay minerals // Proceedings of the 6th International Symposium on Cement and Concrete and Canmet/A ci International Symposium on Concrete Technology for Sustainable Development (Volume 2) [C]. *Cement Branch of Chinese Ceramic Society, Xi'an, China*: 1560–1566.
Suter, J.L. et al. 2009. Computer simulation on study of the materials properties of intercalated and exfoliated poly(ethylene)glycol clay nanocomposites [J]. *Soft Material*, 5(9): 2239–2251.

Hydraulic Engineering IV – Xie (Ed.)
© 2016 Taylor & Francis Group, London, ISBN 978-1-138-02948-4

Simulation evaluation and optimization of road intersection in Lishui City

Y.M. Du & J.P. Wu
Department of Civil Engineering, Tsinghua University, Beijing, China

Z.B. Chen
Lishui Public Security Bureau, Lishui, China

J.X. Wei & H.T. Zhang
Traffic Police Detachment of Lishui Public Security Bureau, Lishui, China

ABSTRACT: Intersection is one of the key parts of an urban road network. Reasonable design of intersection is important in reducing the traffic congestion and improving the road-network capacity. This paper investigates the intersection of Liyang-Kuocang in Lishui City and analyzes its current traffic situations. Four modification scenarios are proposed from the perspective of traffic organization, traffic channelization, road reconstruction, and signal optimization on the basis of its prominent traffic problems. The current situation and the four scenarios are respectively evaluated and analyzed by utilizing simulation technology. The simulation results show that the scenario of signal-time optimization based on traffic channelization is the best, which can shorten the travel time and ease congestion of intersection effectively.

1 INTRODUCTION

As urbanization took place in recent years in China, traffic congestion has become an obstacle for urbanization. This issue in Lishui has not been as serious as that in large cities. However, traffic congestion has emerged in narrow streets in specific regions at specific time period in the old town as the city develops. On the point of joint and diversion of traffic, intersection is the key point to organize traffic (Li, 2012). Frequent conflicts at intersection points are very likely to lower the efficiency of traffic. These conflicts not only present hazards to cyclists' safety but also decrease the level of service for traffic (Du et al., 2015; Yan et al., 2010). Therefore, a reasonable design of intersection is important for reducing the traffic congestion and improving the road-network capacity.

As one of the most important entry–exit places, traffic flow at Liyang-Kuocang is very high. Because of being restricted by road conditions and outside environment, the intersection experiences frequent congestion during rush hours. Based on analysis of status-quo of traffic condition in the Liyang-Kuocang intersection, this article goes deep into investigation and analysis of specific traffic issues and makes the solution to organization of traffic, road channelization, road reconstruction, and signal timing. Simultaneously, analysis and traffic simulation are used to make the solution more specific, accurate, and reliable.

2 ANALYSIS OF THE TRAFFIC SITUATION AT THE INTERSECTION

Liyang-Kuocang is situated in the western part of the Lishui City where residential houses, stores, markets, and schools are located nearby. The traffic flow is high. The location of Liyang-Kuocang is marked as A in Fig. 1.

2.1 Heavy and unbalanced traffic flow

The traffic flow in Liyang-Kuocang is very high. The traffic flow is the maximum on the entrance on the west and the east sides nearly borders. Left-turn traffic flows at the south and north entrances are heavy, and the queue length is long.

2.2 Two adjacent intersections are too close

Because the distance between the intersection of Kuocang-Dengta and Liyang-Kuocang is less than 200 m, the line of cars waiting at Liyang-Kuocang intersection during rush hours sometimes extends to the adjacent intersection of Kuocang-Dengta. Therefore, the joint control of two traffic lights at that intersection will have a huge effect on the efficiency of passing by.

2.3 Unbalanced traffic flow

Vehicles at the Kuocang-Jiefang intersection (marked as G in Fig. 1) can reach the Liyang-Huanchengnan intersection through two routes: H–I–J and F–E–A–B–C. The length of the two routes is 3 km. However, most vehicles like to choose to use the route G–F–E–A–B–C–D to enter or exit downtown, and due to that congestion takes place at GA, AD, and intersection A. On the contrary, the traffic flow at HJ and JD is less. Actually, the shortage of road capacity leads to unevenness in traffic.

2.4 Conflict between nonmotorized vehicles and automobiles

Though regional intersections have some infrastructure, traffic flow of automobiles and nonmotorized vehicles is very high. Pedestrians don't get enough time to cross the road. A lot of nonmotorized vehicles occupy the driveway, and optional lane changing takes place frequently.

Figure 1. Location of Liyang-Kuocang intersection.

Table 1. Delay and LOS of each approach.

	East entrance	West entrance	South entrance	North entrance
Delay (s)	40	31	61	50
LOS	C	B	D	D

3 BENEFIT EVALUATION ANALYSIS AND SIMULATION OF THE TRAFFIC SITUATION

Field data collection is conducted and that for the traffic flow, signal time, and channelization are obtained. The current situation is simulated on the basis of the data collected and illustrated in Fig. 2.

Delay and level of service are calculated on the basis of the simulation results. As indicated in chart 1, the LOS of the south and north entrance is D, which means that traffic density is high, and speed of automobiles is restricted. When the traffic light is red, the queuing length is long, and drivers face discomfort and inconvenience.

4 SIMULATION AND ANALYSIS OF AN ALTERATION SOLUTION TO THE INTERSECTION

Traffic diversion, optimization of lanes, prohibition of left turn, and coordinated signal control are proposed to solve the present problem. The best solution needs to be selected among these four solutions according to the simulation results.

4.1 Solutions description

1. Solution 1: traffic diversion
    The vehicles at the intersection of Kuocang-Dengtajie (marked as E in Fig. 1), the intersection of Kuocang-Jiguang (marked as F in Fig. 1), the intersection of Kuocang-Jiefang (marked as G in Fig. 1), and the intersection of Kuocang-Lutang (marked as H in Fig. 1) are controlled with necessary guidance and diverted to route H–I–J using VMS to lower traffic flow on the route F–E–A–B–C. It alleviates congestion on the left-turn lane on the south entrance of Liyang-Kuocang. This solution is based on the assumption that 40% of the flow would be guided to H–I–J.
2. Solution 2: road channelization
    The original right-turn lane at the south entrance of Liyang-Kuocang intersection is changed to left- and right-turn lane. The channelization is demonstrated in Fig. 3.
    The optimized signal time and the green split are calculated using the Webster Algorithm based on the new channelization. $y_1 = 0.2869$, $y_2 = 0.1328$, $y_3 = 0.2172$, $y_4 = 0.1779$, $Y = 0.814$. The optimized signal time is listed in Fig. 4.

Figure 2.   Current situation simulation.

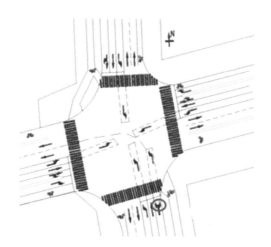

Figure 3. Channelization of Liyang-Kuocang intersection.

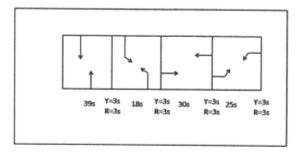

Figure 4. Optimized signal timing.

3. Solution 3: prohibiting left turn at the north entrance

As traffic flow is high during rush hours, the number of cars making left turn is comparatively low. After left turn was prohibited, passing time for vehicles going through the south entrance increased. Then, one of the through-lanes at the south entrance can be used as through-left turn lane, which can alleviate the pressure of left-turn traffic. The channelization based on new solution is displayed in Fig. 5.

4. Coordinated control

Coordinated control is adopted in two adjacent intersections: Kuocang-Dengta (E in Fig. 1) and Liyang-Kuocang (A in Fig. 1). These two intersections are controlled on the basis of Liyang-Kuocang. The signal timing is listed in Table 2.

4.2 *Evaluation*

Delay and travel time of both current and four solutions are compared and shown in Fig. 6.

1. After diversion is used, left-turn traffic flow at the south entrance will decrease. The average delay decreases by 17%, and the average travel time decreases by 12%. Diversion can alleviate traffic congestion, but the result is still determined by the ratio of vehicles that can be diverged.
2. In solution 2, the channelization will increase the number of left-turn lanes. From the angle of simulation, this solution increases the capacity of left-turn traffic. The average delay decreases by 13%. However, inaccuracy may be produced because the driver is not

128

Table 2. Signal time of coordinated control.

Intersection	Phase	Green Time (s)	Amber Time (s)	Red Time (s)	Offset (s)
Liyang-Kuocang	Through traffic at the south and the north entrance	74	3	3	88
	Through and left-turn at the north entrance	16	3	3	
	East–west	30	3	3	

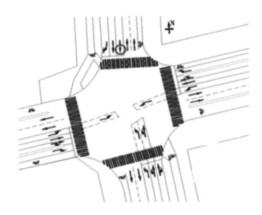

Figure 5. Prohibiting left turn at the north entrance.

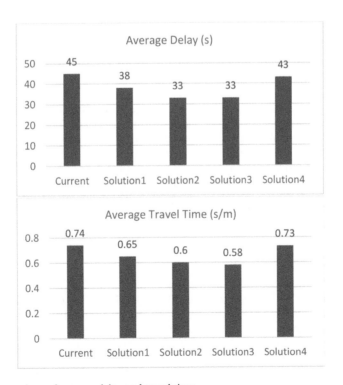

Figure 6. Comparison of average delay and travel time.

Table 3. Comparison of the level of service among current and solutions.

	Current	Solution 1	Solution 2	Solution 3	Solution 4	Suggested solution
LOS	C	B	B	B	C	B

Table 4. Comparison of LOS of solution 2 at each of the entrances.

LOS	East	West	South	North
Current	C	B	D	D
Suggested solution	B	A	B	B

habitual making left turn on the right lane. The radius must be considered as the vehicles make left turn on that lane. Otherwise, conflict may occur between left-turn vehicles at the north entrance and cars from the south entrance.

3. Prohibition of left turn at the north entrance will increase passing time of through and left-turn traffic at the south entrance, thus alleviating the pressure at the south entrance. Vehicles need a detour to make left turn at the north entrance which leads the capacity to decrease by 20%. The average travel time and delay of the whole intersection increased. Problems may occur if left turn is prohibited at the north entrance, and channelization is used at the south. Disorder may be created if vehicles choose different routes inviting congestion.

4. Through and left-turn traffic flow at the south entrance of Liyang-Kuocang are high. Coordinated traffic signal control on the two adjacent intersections does not result in significant change in improvement of the congestion on the intersection of Liyang-Kuocang.

4.3 Suggestion of solution

In accordance with the data collected from simulation, the level of service of current and solutions are shown in Table 3. Four solutions proposed in this paper will be instrumental in improving the traffic situation. Taking into account the angle of difficulty of enforcement and its impact on regional traffic, intersection channelization is a better option. As indicated through the simulation result, the average travel time decreases by 19%, and the level of congestion decreases significantly. The average delay at the south and the north entrances decreases by 46% and 26%, respectively. The level of service, which is shown in Table 4, will increase to level B from level C if this solution is applied.

Based on the analysis above, optimized signal timing needs to be applied on the basis of the channelization of Liyang-Kuocang intersection. This solution poses less impact on the traffic in other directions and regions. But it takes time for drivers to adapt to the new channelization. The traffic sign is required at specific locations.

5 CONCLUSION

The paper focuses on the solution to the traffic congestion at the Liyang-Kuocang intersection in the city of Lishui. Through investigating and analyzing the traffic situation at that intersection, and based on the potential of the present traffic, four solutions are proposed. The current situation along with the four solutions are simulated and evaluated. The result indicates that the best option among the four solutions is optimization of signal timing on the basis of channelization. This solution can effectively shorten average travel time, alleviate congestion, and improve road conditions. Thus, it's scientific and applicable.

ACKNOWLEDGMENTS

The research reported in this paper was supported by Open Foundation of smart-city research center of Hangzhou Dianzi University, smart-city research center of Zhejiang province and funded by The Ministry of Science and Technology of the People's Republic of China (Project No. 2012 AA063303).

REFERENCES

Du, Yiman & Wu, Jianping & Qi, Geqi & Jia, Yuhan. 2015. Simulation study of bicycle multi-phase crossing at intersections. ICE Transport. 168(5), pp:457–465.

Liu, Peihua & Yu, Quan & Liu, Jinguang. 2009. Research on the Channelization and coordinated signal control [J]. Journal of Transport Information and Safety, (3): 24–27.

Ren, Aizhi & Zhou, Weixian. 2007. Exploration of optimized design of city intersection [J]. Journal of Zhejiang transportation college, 7(4):24–26.

Su, Qiangqiang & Ba, Xingqiang. 2010. Simulation evaluation and optimization of intersection in Winter in Herbin [J]. Journal of highway and transportation research and development, 6(10):269–272.

Yan, Ruixue & Hu, Yongju. 2010. Simulation and optimization design of signalized intersection based on VISSIM software[J].Transport standardization, (1):90–93.

Zhang, Li. 2012. Optimization of urban road intersection based on VISSIM simulation software [J]. Traffic Engineering, 9(4):135–138.

Zhou, Tongme & Yang, Jihui. 2005. Research on the solution of urban traffic congestion [J].Journal of Chinese people's public security university: Science and technology, 11(1):98–100.

Hydraulic Engineering IV – Xie (Ed.)
© 2016 Taylor & Francis Group, London, ISBN 978-1-138-02948-4

Change in tunnel inflow due to groundwater level drawdown during excavation

J.S. Moon, A.Q. Zheng, J.Y. Cho & W.J. Park
School of Architectural, Civil, Environmental and Energy Engineering, Kyungpook National University, Deagu, South Korea

ABSTRACT: Groundwater may cause collapse of tunnels as well as ground subsidence during tunneling. Generally, analytical solution used for estimating the groundwater inflow rate does not consider groundwater level drawdown and can overestimate the inflow rate during design phase. In this study, parametric analysis was performed to investigate the effect of groundwater level drawdown.

1 INTRODUCTION

During tunnel excavation, groundwater may lead to collapse of tunnel as well as ground subsidence (Shin, 2005; Rim, 2008). Therefore, an adequate assessment of the groundwater inflow rate and pore water pressure distribution around a tunnel is essential for safety of tunnel construction.

An image tunnel method proposed by Harr (1962) and Goodman et al. (1965) is generally used for estimating the groundwater inflow rate into a tunnel. As shown in Figure 1, Image tunnel method places an image tunnel as a water source at the same distance from the initial phreatic line as the original tunnel. By placing the image source tunnel, the complicated semi-infinite boundary problem can be converted into a relatively simple infinite boundary problem; from which, Eq. (1) can be derived.

$$Q = \left(-\frac{2\pi H \cdot \kappa}{\ln\left[1+\left(\dfrac{2H}{a}\right)^2\right]^{0.5}} \right) \approx -\frac{2\pi H \cdot \kappa}{\ln\left(\dfrac{2H}{a}\right)} \tag{1}$$

where Q is the groundwater inflow rate into a unit length of tunnel; k is the equivalent hydraulic conductivity of a surrounding tunnel; a is the radius of tunnel; and H is the initial hydraulic head from the center of tunnel.

However, analytical solution, used for estimating groundwater inflow rate, does not consider groundwater level drawdown and can overestimate the inflow rate during design phase (Moon, 2013).

In this study, parametric analysis, using a numerical mode, was performed to investigate the effect of drawdown at the groundwater level on hydraulic behavior of ground round a tunnel and transient flow rate into a tunnel.

2 NUMERICAL MODELING

Figure 2 shows the schematic outline for numerical simulation. A tunnel with a diameter of 3 m is located at various depths of GL −15 to −150 m in numerical models. The distance

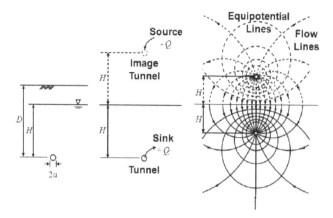

Figure 1.   Image tunnel method (Moon & Fernandez, 2010).

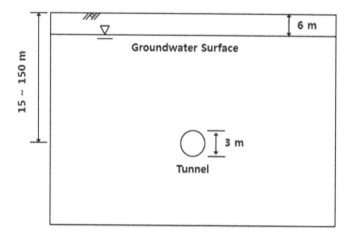

Figure 2.   Numerical model.

Table 1.   Scenarios of parametric study.

	Tunnel depth (m)	Initial groundwater level (m)	Hydraulic conductivity (cm/s)	Groundwater level drawdown
Case 1	GL −15−150	GL −6	10^{-4}	Not allowed
Case 2	GL −15−150	GL −6	10^{-4}	Allowed

of vertical and bottom boundaries from a tunnel are determined to minimize the impact of hydraulic boundary condition.

The groundwater is located at GL −6 m, and the coefficient of hydraulic conductivity of ground k is assumed 10^{-4} cm/s. Table 1 summarizes the cases of parametric study to investigate the effect of groundwater level drawdown during tunnel excavation.

3 PARAMETRIC STUDY RESULT

Figure 3 shows the difference of groundwater inflow rate between two cases in Table 1. It was found that Case 1, ignoring the drawdown at the groundwater level, corresponds well to

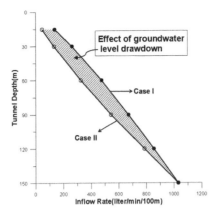

Figure 3.   Steady inflow by tunnel depth.

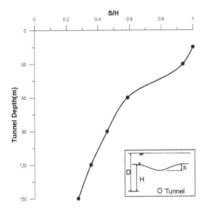

Figure 4.   Groundwater inflow by tunnel depth.

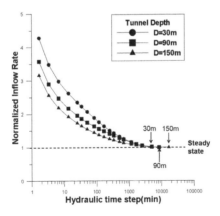

Figure 5.   Inflow rate of tunnel depth according to time.

the estimated values from Eq. (1). The mirror image method was proposed by Harr (1962) and Goodman et al. (1965). As mentioned in the previous section, the image tunnel method overestimates the inflow rate than Case 2 which allows groundwater level drawdown. Thus, the shaded area in Figure 4 represents the effect of groundwater level drawdown.

135

Figure 4 shows the degree of groundwater level drawdown with respect to the tunnel depth. It was found that shallow tunnel excavation causes larger groundwater level drawdown and, consequently, triggers larger reduction of groundwater inflow rate as shown in Figures 3 and 4.

On the other hand, the degree of groundwater level drawdown is relatively small, and the reduction of groundwater inflow rate into a tunnel is negligible.

Figure 5 shows the normalized inflow rate change with hydraulic time step with respect to the steady-state inflow rate. It is found that shallower tunnel experiences a higher initial flow rate (Flush flow), and the inflow reduces rapidly with time, and the flow pattern changes from vertical to horizontal as the groundwater level drawdown occurs.

4   CONCLUSIONS

The parametric study was incorporated in this study to investigate the effect of groundwater level drawdown. It was found that the effect of groundwater level drawdown cannot be ignored for shallow tunnel, and the generally used analytical solution (image tunnel method) can overestimate inflow rate into a tunnel. Therefore, the analytical solution should be revised to consider the degree of groundwater level drawdown. In this study, the reduction of permeability with depth was not considered, and thus, further study should be done to investigate the effect of permeability reduction on hydraulic behavior of ground.

ACKNOWLEDGMENT

This research was supported by a grant (16 AWMP-B079625–03) from Water Management Research Program funded by Ministry of Land, Infrastructure and Transport of Korean Government.

REFERENCES

Goodman, R.E. & Moye, D.G, Van S.A. and Javandel, I. 1965. Groundwater inflows during tunnel riving, *Eng. Geol.*: Vol. 2, No. 1, pp.39–56.

Harr, M.E. 1962. Groundwater and Seepage, *Chap.*: 10, pp. 249–264.

Moon, J.S. & Fernandez. G. 2010. Effect of Excavation-Induced Groundwater Level Drawdown onTunnel Inflow in a Jointed Rock Mass, *Engineering Geology*.: 110(2010) 33–42.

Moon, J.S. 2013. Ground inflow rate estimation considering excavation-induced permeability reduction in the vicinity of a tunnel, *J of Korean Tunn Undergr Sp Assoc.*: 15(3) pp. 333–344.

Rim, H.G. 2008. Predictive analysis of hydraulic change of ground water by tunnel construction using mathematical modeling-In the case of Gyeryong tunnel, *University of Kwangwoon, Master Thesis.*: p.174.

Shin, J.H. 2005. Behaviour of Leaking Tunnels under Unconfined Flow Condition, *Jour. of the KGS.*: Vol, 21, No. 7. September 2005, pp. 43–54.

Hydraulic Engineering IV – Xie (Ed.)
© 2016 Taylor & Francis Group, London, ISBN 978-1-138-02948-4

An experimental study on the bonding and shearing performance of new Modified Polymer Concrete (MPC) with high strength and super lightweight using fly ash

G.W. Liu, X.L. Chen, S.P. Li & F. Gao
School of Transportation Science and Engineering, Bridge and Tunnel Engineering, Harbin Institute of Technology, Harbin, China

G. Yang & Z.C. Hu
Jinzhou Municipal Design Institute, Liaoning Province, China

ABSTRACT: In recent years, because of the large amount of heavy traffic and increase of bridge service life, the damage and disease of highway bridges have become more and more serious in China. In order to meet the needs of the society, we developed a new modified polymer concrete, which has super lightweight and high bending strength. The experimental research results show that two kinds of material tests, such as tensile bonding test and direct shearing test, were carried out. The main results were obtained as follows: (1) the bonding and shearing strengths of this new modified concrete are both larger than that of normal polymer concrete; (2) the bonding and shearing strengths of this new modified concrete depend on that of substrate concrete, since the damage exists in the substrate concrete rather than the interface between the new material and substrate concrete; (3) both bonding and direct shearing strengths meet the requirements of the current specification in China.

1 INTRODUCTION

1.1 *Background*

Because of the large amount of heavy traffic and increase of bridge service life, the damage and disease of highway bridges have become more and more serious in China. Currently, under the condition of shortage of the finance, we face the difficulties in taking effective measures to improve the service life of the bridge.

In recent years, some research workers have developed other kinds of fabric materials similar to Modified Polymer Concrete (MPC) to strengthen reinforced concrete beams. Although those fabric materials could avoid most of the disadvantages, they cannot be used for simultaneous flexural and shear strengthening of beams, because fibers in the materials are in only one direction. Thus, it was necessary to find a new kind of material that could be applied in the engineering.

In order to meet the needs of the society, we have developed a new MPC material, which is made of polyol, polyisocyanate, fly ash, and several modifying agents such as fire retardant, silicon oil, and catalyzer, having super lightweight and high bending strength. The experimental research results show that it not only has many excellent mechanical properties, but also has the advantages of simple and short-time construction process. Unfortunately, there are little reports on bonding and direct shearing characteristics between the substrate reinforced concrete and this new MPC.

1.2 *Outline of experiment*

In the above-mentioned depiction, the focus of this study is to find the bonding and direct shearing behaviors between the MPC material and substrate concrete. Two kinds of experiments, such as tensile test and direct shearing test, were carried out. The main results were obtained as follows: (1) The bonding and shearing strengths of this new modified concrete are both larger than that of normal polymer concrete; (2) the bonding and shearing strengths of this new modified concrete depend on that of substrate concrete, since the damage exists in the substrate concrete rather than the interface between the new material and substrate concrete; (3) both bonding and direct shearing strengths meet the requirements of the current specification in China; (4) the MPC material cannot only be used for seismic design, but can also be used as the rapidly strengthening material of reinforced concrete bridges, due to its large capacities of transformation, remarkable lightweight, and high strength.

2 MATERIALS

2.1 *Chemical components of fly ash*

In this study, the fly ash was used as the main chemical components of this new material. The main components of the fly ash were SiO_2 and Al_2O_3, comprising 70–80% overall. The components of used fly ash are listed in Table 1.

2.2 *Polyurethane components*

The main component of the MPC is polyurethane, which is an excellent polymer elastic material and has high-adhesive properties. The chemical composition of polyurethane is shown in Table 2. Polyether and polyisocyanate were the main raw materials in the mix design, which was used to develop a series of polyurethane filler composites by measuring the density of the new material (MPC). These raw materials can produce different series of polyurethane cement, depending on composition ratio, which gives a flexibility of different densities and mechanical properties. In this research, the target densities (t/m³) were listed in Table 3. The hardness value of this material ranges from 10 to 100 (IRHD). All

Table 1. Chemical components of fly ash.

Chemical components	Weight ratio (%)
SiO_2	55–69
Al_2O_3	20–38
Fe_2O_3	1.5–5.3
CaO	2.0–6.4
MgO	0.2–2.8

Table 2. Main chemical components of polyurethane.

Chemical components	Weight ratio (%)
Polyether	47–49
Fire retardant	0.1–0.5
Silicon oil	0.5–1.5
Catalyzer	0.1–0.5
Polyisocyanate	40–60

components of the polyurethane are liquid materials and were used in this research mixed with fly ashes.

2.3 MPC composites sample

First, the steel molds are to be carefully prepared; in Table 3, the components and ratios of materials are presented. Second, the raw materials of polyurethane and cement are added together, and all materials are mixed for about 3–5 min and then poured into a steel mold. The curing time may vary depending on the amount of contamination. In this experiment, the ambient temperature was around 20–25°C, foam was terminated after 1 h, and the steel mold was removed after 24 h. All the samples were tested after 24 h.

3 EXPERIMENTAL PROCESSES

The first task in the experimental work was to investigate the mechanical properties of the new material (MPC), providing fundamental basic data to proceed with inspecting the bonding behavior of this material when it is used as a kind of strengthening material for bridge or other structural elements. This preliminary study was extended to include the direct bonding test of MPC material to predict the tensile bonding strength properties, which is significant to evaluate the bonding behavior when the structure is under different loading conditions, especially when using these materials for strengthening structure elements (beams, columns, piers, etc.) to predict the slip condition if the under tensile stress exists.

We choose five samples to test and predict the direct bonding stress between the concrete and MPC material. Before the test, the concrete surfaces are cleaned carefully, and any dust, if present, is removed. We kept surfaces as smooth as possible. Then, we put a plastic tube pipe (diameter is 50 mm) on the concrete cube surface (75 × 75 × 75 mm). Steel shaft or bar is centered in the middle area of the MPC material, and it is placed in such a way that it touches the concrete surface. After being sure that the steel shaft was vertical and perpendicular to the concrete surface plane, the MPC material is poured into plastic tube. The reason of doing this is that we must avoid any other inducing stresses during the loading test and insure only one direct tension is applying on samples. Figure 1 shows the test device and sample dimension details. The tensile stress is calculated by dividing the maximum tension force over the net contact area between the concrete and MPC. Figure 2 shows the specimen test processes.

As is shown in Figure 3, we selected Z-shaped specimens as the form of the direct shear test. The thickness of the specimen is 150 mm. We made 10 specimens, of which 5 are ordinary concrete specimens, and the rest are the combination of concrete and MPC material. Similar to the tensile bonding strength experiment, the MPC material is kept at the center in the middle area and is in contact with the concrete. Test pieces are needed to be maintained for about 1 month. Then the universal press machine is used in the structural laboratory to carry out the direct shear test.

Table 3. Mix ratio (%) of the new material for all cases.

Components	Case-1 (0.4 t/m³)	Case-2 (0.6 t/m³)	Case-3 (0.8 t/m³)	Case-4 (1.0 t/m³)	Case-5 (1.2 t/m³)	Case-6 (1.4 t/m³)	Case-7 (1.6 t/m³)
Polyether	20–30	20–30	20–30	20–30	20–30	20–30	20–30
Silicon oil	0.7	0.6	0.5	0.4	0.3	0.2	0.1
Catalyzer	0.30	0.25	0.20	0.15	0.10	0.05	0
Fire retardant	0.15	0.15	0.15	0.15	0.15	0.15	0.15
Polyisocyanate	20–30	20–30	20–30	20–30	20–30	20–30	20–30
Fly ash	40–60	40–60	40–60	40–60	40–60	40–60	40–60

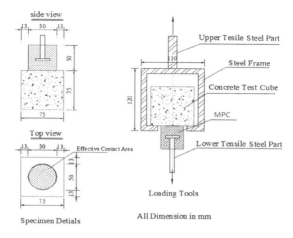

Figure 1. The direct tension bonding test devices and sample details.

Figure 2. Direct tensile test.

Figure 3. Direct shearing test.

4 EXPERIMENTAL RESULTS

4.1 *Tensile bonding strength*

The predicted tensile bonding strength between the MPC material and concrete was higher than concrete bonding materials itself. Table 4 displays the data of the tensile bonding stress.

Table 4. Tensile bonding strength of MPC with concrete surface.

Item	Diameter (mm)	Contact area (mm²)	Tension force (kN)	Bonding strength (MPa)	Average (MPa)
T01	50	1962.5	7.94	4.05	
T02	50	1962.5	7.78	3.96	
T03	50	1962.5	7.86	4.01	4.03
T04	50	1962.5	8.01	4.08	
T05	50	1962.5	7.98	4.07	

Figure 4. The bonding failure of MPC and concrete interface.

Table 5. Direct shearing strength of MPC with concrete surface.

Item	Sizes of L (mm)	Sizes of T (mm)	Contact area (mm²)	Pressure force (kN)	Shearing strength (MPa)	Average (MPa)
S01	150	50	15000	65.3	4.35	
S02	150	50	15000	65.5	4.37	
S03	150	50	15000	66.0	4.40	4.37
S04	150	50	15000	65.8	4.39	
S05	150	50	15000	64.9	4.33	

Figure 4 shows the phenomenon of failure mode of bonding between concrete and MPC material. These experimental results proved that the failure comes from the concrete matrix rather than the interface between MPC and concrete, which indicated that this new material (MPC) has an excellent bonding and adhesive properties.

4.2 Direct shear bonding strength

Table 5 describes the results of the direct shearing test. We know that the MPC material can significantly improve the direct shear bonding strength of the concrete from the table.

5 CONCLUSIONS

This study proposed a new type of material (MPC) that can be used in many different construction fields, especially in strengthening bridges, beams, columns, piers, and other structure elements. The main results of this research are summarized as follows:

1. The bonding and shearing strengths of this new modified concrete are both larger than that of normal polymer concrete.
2. The bonding and direct shearing strength of this new modified concrete depend on that of substrate concrete, because the damage exists in the substrate concrete rather than the interface between the new material and substrate concrete.
3. Both bonding and direct shearing strength meet the requirements of the current specification in China.
4. The composite material (MPC) does not require any special expertise to strengthen or retrofit structural elements, and without special techniques. Generally, this material can make the retrofitting of bridge elements more effective, easier, and cheaper compared with traditional repairing materials.

REFERENCES

ACI Committee 548. 548.1R-09. 2009.Guide for the use of polymers in concrete.American Concrete Institute. Page 1–29.

ACI Committee 548. 548.6R–96.1996.Polymer concrete-structural applications.State-of-the-art report. American Concrete Institute.Page:10–23.

Aidoo John, Harries Kent A, Petrou Michael F. Full-scale experimental investigation of repair of reinforced concrete interstate bridge using CFRP materials. J Bridge Eng 2006;11(3):350–8.

Gorninski JP, Dal Molin DC, Kazmierczak CS. Comparative assessment of isophtalic and orthophtalic polyester polymer concrete: different costs, similar mechanical properties and durability. Constr Build Mater 2007;21:546–55

Grace NF, Abdel-Sayed G, Ragheb WF. Strengthening of concrete beams using innovative ductile fiber-reinforced polymer fabric. ACI Struct J2002.Page:672–700.

Guiwei Liu HO. A foundational study on static mechanical characteristics of the super light weight and high strength material using fly-ash. J Soc Mater Sci 2006;55(8):738.

Hussain Haleem K, Lian Zhen Zhang, Gui Wei Liu. 2013.An experimental study on strengthening reinforced concrete T-beams using new material polyurethane-cement (PUC). Constr Build Mater J 2013.40:104–17.

Karbhari VM, Eckel DA.1993. Strengthening of concrete column stubs through resin infused composite wraps. J. Thermoplast Compos Mater.Page:93–107.

Kukacka LE. 1978.Polymer concrete materials for use in geothermal energy processes. Brook haven National Lab. US Department of Energy. Page: 261–284.

Oded, R., 2008. Debonding analysis of fiber-reinforcedpolymer strengthened beams Cohesive zone modeling versus a linear elastic fracture mechanics approach. Eng. Fract. Mech. 75(10): 2842–2859.

Reis JML, Ferreira AJM. 2004.Assessment of fracture properties of epoxy polymer concrete reinforced with short carbon and glass fibers. Constr Build Mater.

Wang, Y.C. and K. Hsu.2009. Design recommendations for the strengthening of reinforced concrete beams with externally bonded composite plates. Compos Struct.88(2): 323–32.

Hydraulic Engineering IV – Xie (Ed.)
© *2016 Taylor & Francis Group, London, ISBN 978-1-138-02948-4*

Vibration mitigation of nonlinear structural system incorporating uncertainties

N.B. Wang & L.Y. Li
School of Civil Engineering, Dalian University of Technology, Dalian, Liaoning, China

ABSTRACT: This paper introduces a novel control method for mitigating the structural nonlinear vibration. The Sliding Mode Control (SMC) is combined with the Unscented Kalman Filter (UKF) observer. In detail, the UKF is used for identifying the uncertain parameters and estimating the structural states. Based on the estimated information, the SMC is used to determine the control law by these states and parameters. The effectiveness of the SMC-UKF is verified through the numerical simulation. The numerical model is a nonlinear 3-story frame structure containing uncertain parameters which will be updated in real time by UKF. Accordingly, the control effect of the SMC-UKF is compared with the response of the structure without control.

1 INTRODUCTION

In recent years, smart materials and structures have led to the construction of increasingly complex structural systems. Due to the uncertainty and nonlinearity of the structure, the control method is suggested to be adjusted real-time. In general, as a state estimator Kalman filter is used to estimate structural states by finite measurable states. Most researches have been done on this problem for linear system. In 1970, Levine W and Athens M put forward an algorithm, LQR-KF, where LQR is combined with Kalman Filter (KF) observer, and proved the feasibility of the algorithm in theory. Liu & Goldsmith (2004) formulated the Kalman filtering problem with partial observation losses and derived the Kalman filter updates with partial observation measurements. The shortcoming of LQR-KF is that this algorithm assumes the parameters of the controlled object is constant and applies only to the linear system. When dealing with structure under severe earthquake, the state equation may become a nonlinear system. Therefore, KF cannot solve this problem of parameter identification.

In order to apply the Kalman filter to the nonlinear case, Bucy and Sunahara proposed and studied Extended Kalman Filter (EKF). When addressing nonlinear problems using EKF, nonlinear system must be linearized. But linearization will produce model error in strong nonlinearity and has heavy operation burden due to the calculation of Jacobin matrix. Ghosh et al (2007) proposed two novel forms of the EKF. These filters are based on variants of the derivative-free Locally Transversal Linearization (LTL) and Multi-step Transversal Linearization (MTrL) schemes. The proposed filters do not need computing Jacobin matrix at any stage. Heavy operation burden is partly solved, but the model error is still unsolvable. Corigliano et al (2004) & Naets et al (2014) put forward a control method where LQR is combined with EKF respectively. Roffel et al (2014) and Szabat & Orlowska-Kowalska (2006, 2008) applied EKF to different structure respectively. To detect and isolate the faults of the current and voltage sensors, He et al (2015) presented a model-based fault diagnosis scheme which is based on EKF. Torres et al (2015) used EKF to present a nonlinear approach to identify the structural parameters of a vertical riser.

Unscented Kalman Filter (UKF) is combined Unscented Transform (UT) with KF. The well-known shortcoming of the EKF, especially in the handling of higher order nonlinearities, is that it requires linearization of the system equation. In recent years, the UKF has been intensively researched for a wide range of estimation purposes. It has become quite popular in lots of engineering areas. In recent study, Sunderhauf et al (2007) and Bisgaard et al (2010) introduced the UKF to the aircraft control system respectively. Umoh (2014) studied the antisynchronization of a generalized multi-wing butterfly attractor design template and used UKF to identify the parameters of the system. In the field of civil engineering, Omrani et al (2013) proposed a new parametric approach where the UKF is used to estimate the Bouc-Wen model parameters considering the inelastic response of individual elements in buildings. Miah et al (2015) used UKF as an experienced estimator which can estimate the uncertain parameters and the immeasurability structural states.

In this work, the Sliding Mode Control is combined with UKF (SMC-UKF) to solve the problem of controlling the structure containing uncertainties. Structural states and uncertain parameters are estimated by the UKF based on limited measured states. Control law derived by SMC contains these estimated states and parameters. Therefore the control law can be updated in real-time by UKF. The SMC-UKF can not only solve the vibration mitigation problem, but also can identify uncertain parameters.

2 SMC-UKF

2.1 State estimation

To describe the general class of systems, consider a nonlinear time-invariant structural system in the functional form.

$$x_k = f(x_{k-1}, u_{k-1}) + w_{k-1} \tag{1}$$

$$y_k = H(x_k, u_k) + v_k \tag{2}$$

where, $f(.)$ is a nonlinear state function; $H(.)$ is an observation equation; w_{k-1} is the process noise vector which is assumed to be Gaussian distributed with $N(0, Q_{k-1})$; v_k is the Gaussian measurement noise vector distributed with $N(0, R_k)$. Structural states are estimated by the UKF according to limited measured states. The detailed analysis of the UKF is summarized in Table 1.

2.2 Parameter identification

In order to solve the problem to control the structure containing unknown parameters, equation (1) and equation (2) must be modified. Assuming that all unknown parameters can be summarized in a parameter vector θ, the augmented state vector is defined as

$$\bar{x} = [x \quad \theta]^T \tag{3}$$

The process and observation equations are as follows:

$$\bar{x}_k = f(\bar{x}_{k-1}, u_{k-1}) + w_{k-1} \tag{4}$$

$$y_k = H(\bar{x}_k, u_k) + v_k \tag{5}$$

When controlled structure contains unknown parameter θ, the process equation must be nonlinear since it includes the bilinear products of the components x and θ. This is the reason why KF cannot identify parameters in this case.

144

Table 1. The general scheme of the UKF algorithm for state and parameter estimation.

― Initialization at time t_0

$$\hat{x}_0^+ = E(\bar{x}_0), \quad P_0^+ = E[(\bar{x}_0 - \hat{x}_0^+)(\bar{x}_0 - \hat{x}_0^+)^T]$$

― At time t_{k-1}
- The unscented transform
  Formulation of the sigma point vector,

$$\hat{x}_{k-1}^{(i)} = \hat{x}_{k-1}^+ + \tilde{x}^{(i)}, \tilde{x}^{(i)} = (\sqrt{nP_{k-1}^+})_i^T, \tilde{x}^{(n+i)} = -(\sqrt{nP_{k-1}^+})_i^T, i = 1,...,2n$$

- Update stage
  1. Propagation of the sigma points through the measurement model,

$$\hat{y}_k^{(i)} = H(\hat{x}_k^i, u_k, t_k)$$

  2. Calculation of the state and covariance priors,

$$\hat{x}_k^- = \frac{1}{2n}\sum_{i=1}^{2n}\hat{x}_k^{(i)}, \quad P_k^- = \frac{1}{2n}\sum_{i=1}^{2n}(\tilde{x}_k^{(i)} - \hat{x}_k^-)(\tilde{x}_k^{(i)} - \hat{x}_k^-)^T + Q_{k-1}$$

- Update stage
  1. Propagation of the sigma points through the measurement model,

$$\hat{y}_k^{(i)} = H(\hat{x}_k^i, u_k, t_k)$$

  2. Calculation of the state and covariance priors,

$$P_y = \frac{1}{2n}\sum_{i=1}^{2n}(\hat{y}_k^{(i)} - \hat{y}_k)(\hat{y}_k^{(i)} - \hat{y}_k)^T + R_k, \quad P_{xy} = \frac{1}{2n}\sum_{i=1}^{2n}(\tilde{x}_k^{(i)} - \hat{x}_k^-)(\hat{y}_k^{(i)} - \hat{y}_k)^T$$

  3. Calculation of Kalman gain and predictions of state(posterior estimates) using the latest observations

$$K_k = P_{xy}(P_y)^{-1}, \hat{x}_k^+ = \hat{x}_k^- + K_k(y_k - \hat{y}_k), P_k^+ = P_K^- - K_K P_y K_k^T$$

2.3 SMC-UKF control law

Error signal is $e = x-x_d$, where x is the displacement vector of structure and x_d is the desired displacement response. In this paper, let $x_d = 0$ for vibration mitigation purpose. The sliding mode control law is defined according to the test model and the control law should guarantee that the system satisfies certain conditions. The conditions are: (1) the sliding mode exists; (2) the sliding mode surface can be reached; (3) the sliding mode motion of the system is stable.

In this study, the sliding mode surface is defined as:

$$S = C_c e + \dot{e}, C_c = diag(c_1, c_2, c_3) \tag{6}$$

where, $c_i > 0$ (i = 1, 2, 3) are real numbers.
The Lyapunov function can be defined as

$$V = \frac{1}{2}S^T MS \tag{7}$$

where M is the mass matrix. Then, the derivative of V can be calculated as

$$\dot{V} = S^T M\dot{S} = S^T(MC_c\dot{x} - C\dot{x} - F(x) - u - M_e a_g) \tag{8}$$

Control law u is defined as

$$u = MC_c\dot{x} - C\dot{x} - F(x) + m(D+\Gamma)\text{sgn}(S) \tag{9}$$

145

where sgn(.) is a sign function, D is the upper bound of the external excitation. $\Gamma = \text{diag}$ $(\gamma_1, \gamma_2, \gamma_3)(\gamma_i > 0)$. Substituting equation (9) into equation (8), the derivative of V can be rewritten as

$$\dot{V} = S^T(-M_e a_g - mD\text{sgn}(S) - m\Gamma\text{sgn}(S)) \leq m\Gamma|S| \leq 0 \qquad (10)$$

According to Lyapunov stability theory, the designed control law can guarantee the controlled system is stable. And because $c_i > 0$, in equation (6) we can obtain

$$e \to 0 \qquad (11)$$

Therefore, the control law determined by equation (9) can effectively control the structural response under seismic excitation.

3 TEST MODEL—3-STORY SHEAR FRAME

The simulation model is a 3-story shear frame structure and the parameters of the model are presented in Table 2. The nonlinear restoring forces of the structure are assumed to be cubic nonlinear functions. The physical model of the structure is defined as

$$M\ddot{x} + C\dot{x} + F(x) + u = -M_e a_g \qquad (11)$$

$$\text{where } M = m\begin{bmatrix} 1 & & \\ 1 & 1 & \\ 1 & 1 & 1 \end{bmatrix}, C = \begin{bmatrix} c_1 & -c_2 & \\ c_2 & -c_3 \\ & c_3 \end{bmatrix}, F(x) = \begin{bmatrix} k_1 x_1^3 - k_2 x_2^3 \\ k_2 x_2^3 - k_3 x_3^3 \\ k_3 x_3^3 \end{bmatrix}, M_e = m\begin{bmatrix} 1 \\ 1 \\ 1 \end{bmatrix}.$$

M and C are the mass and damping matrices respectively. $F(x)$ is the restoring force of the structure. u is the applied control force. ag is the earthquake acceleration. x is the inter-story displacement vector.

4 SIMULATIONS AND RESULTS

In this section the UKF will be used to estimate states and identify parameters. The stiffness of the model is assumed to be unknown and those parameters are updated online according to the measured acceleration signal. After the estimation of the useful information, the SMC will be used to control the vibration of the nonlinear structure. The flow chart of the whole system is shown in Figure 1.

4.1 *Control law design*

In order to simulate the actual situation, the output bound is added to the control force that is determined by equation (8) and the absolute value is set to be 5e5 N. It is well known that

Table 2. The parameters of the numerical simulation model.

parameters			
storey	m_i(t)	k_i (10^4 kN/m)	c_i (kN*s/m)
1	345.6	9.315	545
2		7.605	445
3		6.165	359

146

the inherent problem of SMC is the chattering phenomenon. In order to eliminate the chattering phenomenon, sgn(.) is replaced by atan(.) in equation (8).

4.2 State estimation and parameter identification

The El Centro earthquake multiplied by a constant 2.868 is used as seismic excitation. The initial assumption of uncertain parameters k_i is set to be half of the actual value. Initial value of state is assumed to be zero. Let process noise $Q = diag(1e\text{-}24\ I_{6,6},\ 1e\text{-}4I_{3,3})$ and observation noise $R = 0.01$. The initial covariance of the states equals to be 1e-24 and the initial covariance of the parameters equals to be 1e2. Therefore $P_0 = diag(1e\text{-}10I_{1,6},\ 1e2I_{1,3})$. The result of parameter identification is shown in Figure 2 and the results of state estimation are shown in Figure 3 and Figure 4.

Through the Figure 3, the identified values approach the actual value in about 3 second. The final error between identified value and actual value is shown in Table 3. The estimated states can effectively track the actual states, as shown in Figure 4 and Figure 5.

4.3 Control effectiveness

The structural inter-story displacement is the most important factor for control purpose. In this study, the control effect of the inter-story displacement is taken into consideration. The control effectiveness is shown in Figures 5–7. The control force is shown in Figure 8. From these Figures, it can be shown that the proposed estimation and control method is effective for vibration suppression of nonlinear structure.

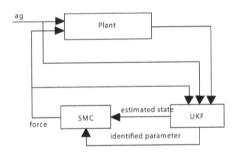

Figure 1.   The flow chart of simulation.

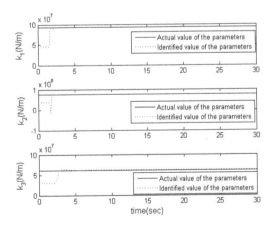

Figure 2.   Convergence of the identified parameters.

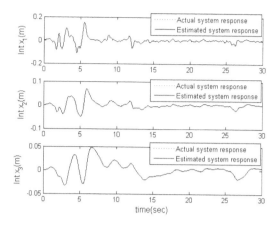

Figure 3.   Comparison of the actual displacements and the estimated displacements.

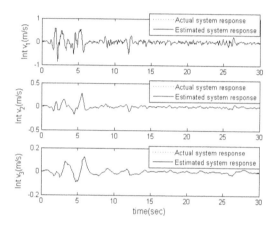

Figure 4.   Comparison of the actual velocities and the estimated velocities.

Table 3.   The identified parameters of the simulation model.

Parameters	Actual (10^7)	Initial assumption (10^7)	Initial error (%)	Identified (10^7)	Final error (%)
k_1	9.315	4.66	50	9.3055	0.1
k_2	7.605	3.80	50	7.6195	0.19
k_3	6.165	3.31	50	6.2312	1.07

5   CONCLUSION

This paper proposed a novel control method of the Sliding Mode Control (SMC) combined with the Unscented Kalman Filter (UKF) observer for mitigating the structural nonlinear vibration. The stiffness of the simulated structure is assumed to be unknown and nonlinear, which is updated in real time based on the measured acceleration signal. The simulation is conducted on a 3-story frame structure with nonlinear stiffness. The results validate that the identified values reach actual value in a short time and the estimated states can effectively track the actual states. The control effectiveness of the proposed SMC-UKF is compared with the response of the structure without control.

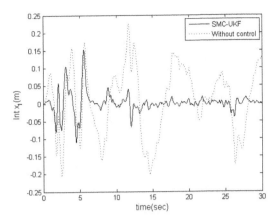

Figure 5. Comparison of 1st floor inter-story displacement with and without control.

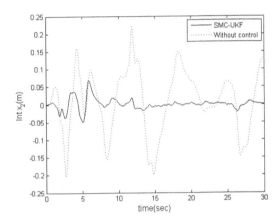

Figure 6. Comparison of 2nd floor inter-story displacement with and without control.

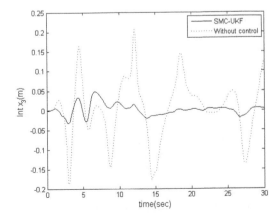

Figure 7. Comparison of 3rd floor inter-story displacement with and without control.

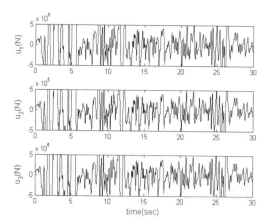

Figure 8. The control force of SMC.

ACKNOWLEDGMENTS

This work was supported in part by the National Science Foundation of China under Award No. 51378093 and No. 91315301.

REFERENCES

Bisgaard M., la Cour-Harbo A., Dimon Bendtsen J. (2010). "Adaptive control system for autonomous helicopter slung load operations." Control Engineering Practice, 18(7), 800–811.

Corigliano A., Mariani S. (2004). "Parameter identification in explicit structural dynamics: performance of the extended Kalman filter." Computer Methods in Applied Mechanics and Engineering, 193(36–38), 3807–3835.

Ghosh S.J., Roy D., Manohar C.S. (2007). "New forms of extended Kalman filter via transversal linearization and applications to structural system identification." Computer Methods in Applied Mechanics and Engineering, 196(49–52), 5063–5083.

He H., Liu Z., Hua Y. (2015). Adaptive Extended Kalman Filter Based Fault Detection and Isolation for a Lithium-Ion Battery Pack. Energy Procedia, 75, 1950–1955.

Liu X, Goldsmith A. (2004). "Kalmen filtering with partial observation losses." Decision and control, 4(43).

Miah M.S., Chatzi E. N., Weber F. (2015). "Semi-active control for vibration mitigation of structural systems incorporating uncertainties." SMART MATERIALS AND STRUCTURES, 24(0550165).

Naets F., Pastorino R., Cuadrado J., et al. (2014). "Online state and input force estimation for multibody models employing extended Kalman filtering." Multibody System Dynamics, 32(3), 317–336.

Omrani R., Hudson R. E., Taciroglu E. (2013). Parametric Identification of Nondegrading Hysteresis in a Laterally and Torsionally Coupled Building Using an Unscented Kalman Filter. JOURNAL OF ENGINEERING MECHANICS-ASCE, 139(4), 452–468.

Roffel A.J., Narasimhan S. (2014). "Extended Kalman filter for modal identification of structures equipped with a pendulum tuned mass damper." Journal of Sound and Vibration, 333(23), 6038–6056.

Sunderhauf N., Lange S., Protzel P. (2007). "Using the Unscented Kalman Filter in mono-SLAM with inverse depth parametrization for autonomous airship control." IEEE, 54–59.

Szabat K., Orlowska-Kowalska T. (2006). "Adaptive control of two-mass system using nonlinear extended Kalman filter." IEEE, 5420–5425.

Szabat K., Orlowska-Kowalska T. (2008). "Application of the Extended Kalman Filter in Advanced Control Structure of a Drive System with Elastic Joint." IEEE, 595–600.

Torres L., Verde C., Vázquez-Hernández O. (2015). Parameter identification of marine risers using Kalman-like observers. Ocean Engineering, 93, 84–97.

Umoh E. A. (2014). Parameter estimation and antisynchronization of multi-wing chaotic flows via adaptive control, 6.

Hydraulic Engineering IV – Xie (Ed.)
© 2016 Taylor & Francis Group, London, ISBN 978-1-138-02948-4

Dynamic response analysis of bridge rectangular pier in deep water

Fan Lei & Yulin Deng
School of Transportation, Wuhan University of Technology, Wuhan, Hubei, China

ABSTRACT: The dynamic response of underwater structures is different from those in air because of interaction between fluid and solid. With a lot of actual examples of deep-water bridges, the dynamic characteristics of rectangular bridge pier in deep water are analyzed on the basis of potential fluid theory. The dynamic response influence of rectangular pier with different depth of water is discussed, under the action of harmonic loads.

1 INTRODUCTION

With the high development of transportation, more and more long-span bridges are under construction throughout the world. The depth of underwater piers is increasing especially in deepwater and sea-crossing bridges. Recent studies have shown that the dynamic response of underwater structures is different from that in air (Lei 2009 & Deng 2014). Therefore, the interaction between structure and fluid should be taken into consideration while analyzing the dynamic response of deepwater pier. It's very difficult to carry out theoretical analysis on fluid–solid interaction problems; only the numerical method can be used in the study (Kai & Wan 2013). It is still difficult to determine the surface contact between fluid and solid, and between fluid scope and constraint condition, which needs to be further studied, while calculating the dynamic response of deep-water bridge. The dynamic characteristics and dynamic response of a rectangular pier of 60 m are studied under harmonic loads.

2 NUMERICAL MODELS

A rectangular pier is chosen to be analyzed. It has a height of 60.0 m, a length of 7.5 m, a width of 4.5 m, and a thickness of 0.8 m. C40 concrete is the main material of rectangular pier. It has a density of 2500 kg/m³, a Young's modulus of 30 GPa, and a Poisson's ratio of 0.2. Water density is 1000 kg/m³. A finite-element model of the pier is shown in Figure 1.

3 DYNAMIC CHARACTERISTICS ANALYSIS OF RECTANGULAR BRIDGE PIER

It is assumed that the water is still, and it has an ideal infinite boundary. The upper surface of water is set to be free surface. All around the water is set to be infinite boundary. The bottom of water and pier is concretion. The pier and water are simulated by three-dimensional solid and fluid elements, respectively. The interface of structure and fluid is coupling interface. The first three natural vibration periods on longitudinal and transverse direction of pier under different depths of water are shown in Tables 1 and 2.

It is shown in Tables 1 and 2 that the first three natural vibration periods of pier increase with the increase in depths of water. The influence of fluid–solid interaction on a pier natural vibration period of the longitudinal direction is greater than that in the transverse direction.

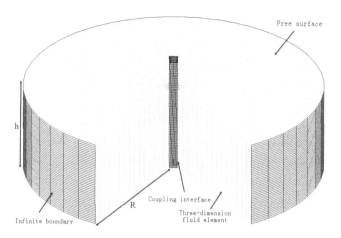

Figure 1. Finite-element model of the pier with water.

Table 1. The first three natural vibration periods with different water depths (longitudinal direction).

Water depth (m)	First vibration period (s)	Second vibration period (s)	Third vibration period (s)
0	1.144	0.197	0.079
10	1.144	0.199	0.082
20	1.150	0.217	0.096
30	1.184	0.251	0.096
40	1.282	0.264	0.104
50	1.472	0.265	0.108
60	1.761	0.295	0.115

Table 2. The first three natural vibration period with different water depth (transverse direction).

Water depth (m)	First vibration period (s)	Second vibration period (s)	Third vibration period (s)
0	0.740	0.130	0.054
10	0.740	0.130	0.054
20	0.742	0.130	0.058
30	0.751	0.132	0.059
40	0.778	0.135	0.061
50	0.835	0.139	0.064
60	0.927	0.143	0.066

Among the first three natural vibration periods, the change of first natural vibration period is significant. The first natural vibration period of the longitudinal direction is increased by 12.05% when the depth of water is 40 m and is increased by 53.90% when the water depth reaches 60 m. Therefore, the depth of water has a great influence on the natural vibration period of pier.

4 HARMONIC RESPONSE ANALYSIS OF RECTANGULAR BRIDGE PIER

Mass proportion load is applied on the pier for simulating the moving of ground. The frequencies of the harmonic load are 4, 2, 1.5, 1, 0.75, and 0.5 Hz to which the corresponding load periods are 0.25, 0.5, 0.67, 1, 1.33, and 2 s, respectively. Viscous fluid is used to simulate water. The dynamic response influence of displacement of pier top, the bending moment, and shearing force of pier bottom of rectangular pier with different depths of water is discussed.

Under the action of longitudinal direction load, the curve of pier top displacement of 60 m rectangular bridge pier with different load periods under different water depths is shown in Figure 2. The curve of bending moment and shearing force of pier bottom of the pier with different water depths are shown in Figure 3. Under different water depths, the time–history curve of pier top displacement with longitudinal direction load periods of 1 and 1.33 s are shown in Figures 4 and 5.

It can be seen from Graphs 2–5 that the variation trends of pier top displacement, pier bottom bending moment, and shearing force are consistent under the action of longitudinal direction load of bridge. The peak value of pier top displacement is negatively related to the water depth when the load period is 1 s. When the load period increases to 1.33 s, the peak value of pier top displacement reaches the maximum value under the water depth of 45 m. The peak value of pier top displacement is the minimum when the water depth is 0 m.

The first natural vibration period of longitudinal direction of the pier is 1.144 s when the water depth is 0 m. The resonance effect and the dynamic response are obvious when the load period is 1 s. When the water depth increases, the natural vibration period keeps away from the load period, and the resonance effect and the dynamic response decrease. That is, the action of fluid–solid interaction reduces the dynamic response of the pier. The resonance effect and the dynamic response are weak when the load period is 1.33 s, for it is far away from the natural vibration period of the pier. With the increase in water-depth, the natural vibration period

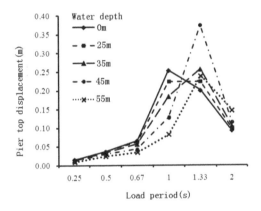

Figure 2.   Pier top displacement under longitudinal direction load.

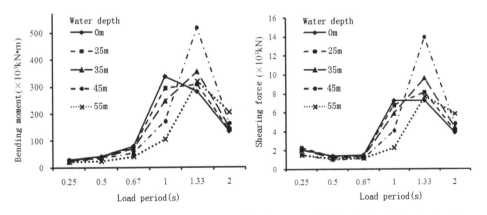

Figure 3.   Bending moment and shearing force of pier bottom under longitudinal direction load.

153

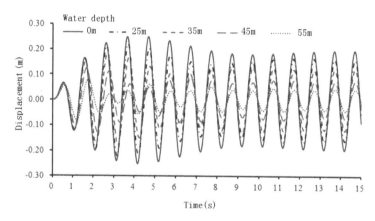

Figure 4. Time–history curve of pier top displacement under longitudinal direction load period of 1 s.

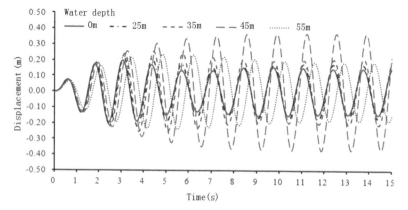

Figure 5. Time–history curve of pier top displacement under longitudinal direction load period of 1.33 s.

becomes approximately 1.33 s, and the resonance effect and the dynamic response increase. That is, the action of fluid–solid interaction increases the dynamic response of the pier.

Under the action of transverse direction load, the maximum pier top displacement with different water depth is shown in Figure 6. The curve of bending moment and shearing force of pier bottom of the pier with different water depths are shown in Figure 7. The time–history curve of pier top displacement with different water depths under the action of transverse direction load periods of 0.67 and 1 s are shown in Figures 8 and 9.

It can be seen from Graphs 6–9 that the variation trends of pier top displacement, pier bottom bending moment, and shearing force are consistent under the action of transverse direction load of bridge. When the load period is 0.67 s, the peak value of pier top displacement with water depth of 0 m is the maximum, and the value is the minimum with the water depth of 55 m. The peak value of pier top displacement is negatively related to the water depth. When the load period increases to 1 s, the peak value of pier top displacement becomes positive, related to the water depth.

The first natural vibration period of transverse direction of the pier is 0.74 s, when the water depth is 0 m. The resonance effect and the dynamic response are strong when the load period is 0.67 s. When the water depth increases, the natural vibration period keeps away from the load period, and the resonance effect and the dynamic response decreases. That is, the action of fluid–solid interaction reduces the dynamic response of the pier. The resonance

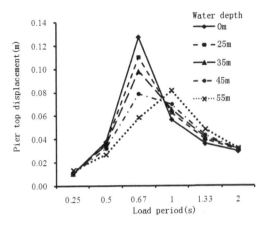

Figure 6. Pier top displacement under transverse direction load.

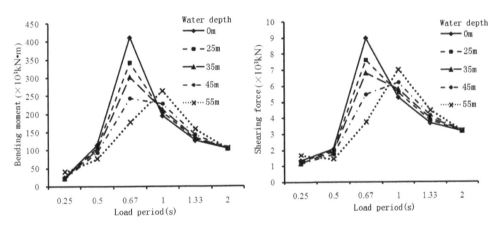

Figure 7. Bending moment and shearing force of pier bottom under transverse direction load.

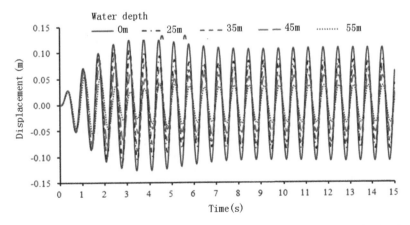

Figure 8. Time–history curve of pier top displacement under transverse direction load period of 0.67 s.

effect and the dynamic response are weak when the load period is 1 s, for the load period is far away from the natural vibration period of the pier. With the increase of water depth, the natural vibration period becomes approximately 1 s, and the resonance effect and the

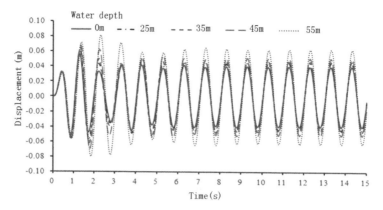

Figure 9.   Time–history curve of pier top displacement under transverse direction load period of 1 s.

dynamic response are increased. That is, the action of fluid–solid interaction increases the dynamic response of the pier.

5   CONCLUSIONS

The dynamic response of a rectangular pier under harmonic loads is studied on the basis of three-dimensional model and viscous fluid. The dynamic response influence of displacement of pier top, the bending moment, and shearing force of pier bottom of rectangular pier with different depth of water is discussed. The action of fluid–solid interaction increases the dynamic response of the pier when the load period is larger than the nature vibration period of the pier. On the contrary, the action of fluid–solid interaction decreases the dynamic response of the pier. When the period of external load is much larger or much less than the nature vibration period, the action of fluid–solid interaction on dynamic response of the pier can be ignored.

ACKNOWLEDGMENTS

This research was supported by the National Nature Science Foundation of China (Grant no. 51378406).

REFERENCES

Deng, Y. L, & Bian, Y. & Lei,F. Earthquake Response Analysis of Long span the flexible submarine pipeline. *Advances in Civil and Industrial Engineering IV*. 2014:1704~1707.

Kai, W. & Najib, B. Experimental and numerical assessment of the three-dimensional modal dynamic response of bridge pile foundations submerged in water. *Journal of bridge engineering.* 2013(18):1032–1041.

Lei, F. & Yang, J.X. & Xie, X.Z.: The Numerical Calculation Research on Dynamic Characteristics of Pipe Conveying Fluid. *Recent Advances in Nonlinear Mechanics2009, Kuala Lumpur*:72–73.

Wan, L.Y. & Qiao L.A. New Added Mass Method for Fluid-structure Interaction Analysis of Deep-water Bridge. *Journal of Civil Engineering*.2013,17(6):1413–1424.

Hydraulic Engineering IV – Xie (Ed.)
© *2016 Taylor & Francis Group, London, ISBN 978-1-138-02948-4*

Study on durability and service life of anchor bolt structure

Lili Zhang
Beijing Polytechnic College, Beijing, China

Qinxi Zhang & Xiaojie Wang
College of Architecture and Civil Engineering, Beijing University of Technology, Beijing, China

ABSTRACT: With a great number of applications in various engineering, the durability and service life of the anchorage structure have become an important subject in the process of the development of anchorage technology. In this paper, using the method of combining indoor experiment with theoretical analysis, the durability and service life of anchorage structure are comprehensively analyzed by using the grey system theory.

1 INTRODUCTION

The bolt supporting technology is a construction technique by placing tension member in the soil or rock, so as to enhance its own strength and stability. Due to its convenient and fast construction, low cost, strong engineering applicability and ability to use in combination with other forms of support, it has good technical and economic advantages, widely used in the rock and soil slope supporting engineering of water conservancy and hydropower, railway and highway, construction, metallurgy and other industries in many countries. Up to now, the bolt structure has been applied in different engineering fields in China for more than 40 years. However, as the permanent support of numerous projects, how much the remaining life of such anchorage structure is and whether it will lose efficacy suddenly or not has become a question that is widely concerned and difficult to answer. Therefore, there is very important and long-term significance to study the durability and service life of bolt structure. In this paper, according to the indoor accelerated corrosion test results of the cement mortar and the anchor, the coupling factor analysis of anchor bolt corrosion is carried out. Besides, based on the variation of the corrosion rate and the bearing capacity of the anchor bolt, the grey prediction analysis of the service life of the bolt is also carried out.

2 EXPERIMENTAL STUDY ON THE CORROSION OF ANCHOR BOLT AND GREY ANALYSIS OF SERVICE LIFE PREDICTION

2.1 *Experimental design*

The specimen uses HRB400 grade steel, and the size is $\Phi 8 \times 40$ mm after cleaning the oil stain of surface with gasoline and acetone. Place the specimen in the following manners:

1. Confined moisture: the specimen is in a closed container hanging above the surface of the water.
2. Permanent immersion: the specimen is in a closed container placed below the surface of the water 5 cm.
3. Alternate wetting and drying: the specimen is alternately located below the surface of the water 5 cm and above the water surface, and each 7d is exchanged for a second time.

It is used to simulate the acid and alkaline environment that respectively mix dilute sulfuric acid, calcium hydroxide and water to the required pH value and permanently immerse the specimens below the surface of the water 5 cm of the container.

2.2 Test result analysis

As shown in Figure 1, in four different corrosion environment of closed wet, permanent immersion, alternating wet and dry, weak acid aqueous solution, the weight loss rate of anchor bolt body in weak acid aqueous solution is the largest, and that in the closed moist environment is the smallest. The reason is that the anodic reaction of the rod body corrosion battery in the weak acidic aqueous solution is mainly hydrogen reduction in the anoxic condition. The escape force produced by the hydrogen evolution makes the formation of the passive film on the surface of the steel bar become difficult.

The weight loss rate under the condition of alternation of wetting and drying is similar to that under the condition of permanent immersion, but the growth rate of weight loss rate is slightly larger than that under the condition of permanent immersion. This is because in the permanent immersion condition, the passivation film on the surface of the steel bar once formed, the further development of corrosion will become very slow.

Under closed wet condition, the weight loss rate is only about 1/5 of the weight loss rate under the condition of permanent immersion and dry wet alternation. This is because the rod surface attached to a layer of air condensed water, the content and conductivity of the water molecules are relatively low, so that the corrosion cell efficiency is not high. In addition, there is no concentration effect of oxygen under the condition.

In no matter what kind of test environment, the weight loss rate of anchor is increased with time, but the increase speed is decreased with time. This is because with the corrosion product thickening in the of the surface of the rod body, the resistance of oxygen or hydrogen ion diffusion to the surface, anodic dissolution and hydrogen escape are all increased.

2.3 Grey prediction analysis

I In practical engineering, the bearing capacity of the steel bar has a direct influence on the safety of the structure, so the loss rate of the bearing capacity of the steel bar is chosen as the failure criterion. Using the safety factor 1.2 that is commonly used in the project, without considering other factors, the section area of the reinforcement is not changed. So as to calculate the threshold value of the bearing capacity loss rate for 16.67%. Thus the corresponding loss rate of steel bar section can be obtained 14.76%.

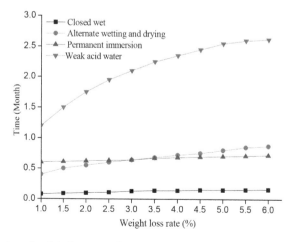

Figure 1.   Distribution of weight loss rate of steel bar with time.

According to the steps of sequence grey prediction, the MATLAB algorithm program is compiled to calculate and test. Sequence grey prediction steps are: ① level ratio test and modeling feasibility judgment; ② data transformation processing; ③ GM (1,1) modeling; ④ test; ⑤ prediction; ⑥ forecast.

In the process of modeling, the residual test method and the level deviation test are adopted to compare the predictive value and the actual value. If the test meets the accuracy requirement, the average value of the positive and negative residuals and is taken into the predicted value after averaging, as the forecast value. Finally, the shift value Q^* is subtracted from the predicted value, which is the final result.

The data sequence of GM (1,1) model under different corrosion conditions is shown in Table 1.

In order to further verify the feasibility of the grey prediction of corrosion test data of anchor bars, compare the predictive value and the actual value, use the former part of the original data sequence as the modeling data, and predict and verify the data of the last period of time. Test results as shown in Table 2, The data of the first 8 groups of the original data sequence is used to predict the data of the latter 3 groups. The results are close, which shows that it is feasible to use the grey prediction to analyze the test results and it has a high degree of confidence.

The weight loss rate to predict the sequence, the parameters, test index and prediction results of the model obtained are shown in Table 3. According to the parameters of GM (1,1) model, we can see that the absolute value of the development coefficient A is much less than 0.3, which illustrates that the model can be used for medium and long term forecasting. At the same time, the average accuracy of the model is more than 0.95 and the maximum deviation of the mean level is only 0.0373, which meet the modeling requirements. As a result, the model is reasonable, and the accuracy of the model is higher.

From the comparison of prediction curve and the original data points (From Fig. 2 to Fig. 5), it can be concluded that the prediction curve passes through the original data points, and then the development trend is consistent with the the latter. Besides, under different conditions, the development situation is obviously different, and it is not a linear change, which is more close to the actual situation.

From the forecast results, according to the selected weight loss rate threshold of 14.76%, the longest life span of the anchor bar is 8.3a under the condition of permanent immersion; the life span under the closed condition is next, which is 8.2a; in weak acid water, the life of anchor bars is the shortest, only 2.4a. This is due to the weak acid environment, the formation of the oxide film passivation layer on the steel surface is easily damaged by acid water, resulting in rapid development of steel corrosion, so life span is greatly reduced.

Table 1. Data sequence used in GM (1,1) model.

Data sequence		Weight loss rate of steel bar		
Serial number	Time/month	Permanent immersion	Alternate wetting and drying	Weak acid water
1	1	0.6609	0.4244	1.1408
2	1.5	0.6614	0.4973	1.4662
3	2	0.6622	0.5630	1.7147
4	2.5	0.6636	0.6214	1.9046
5	3	0.6659	0.6724	2.0496
6	3.5	0.6696	0.7161	2.1604
7	4	0.6758	0.7526	2.2450
8	4.5	0.6859	0.7817	2.3096
9	5	0.7027	0.8036	2.3590
10	5.5	0.7303	0.8181	2.3967
11	6	0.7758	0.8254	2.4255

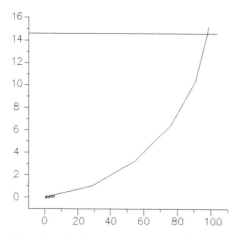

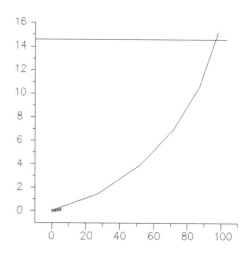

Figure 2. Prediction curves and initial data points in closed wet condition.

Figure 3. Prediction curves and initial data points in permanent immersion condition.

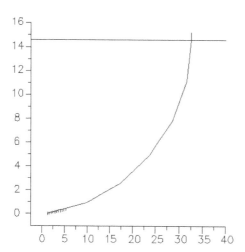

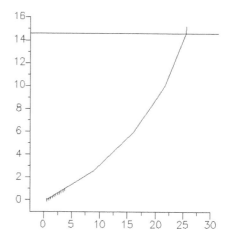

Figure 4. Prediction curves and initial data points in alternate wetting and drying condition.

Figure 5. Prediction curves and initial data points in weak acid water condition.

3   PREDICTION ANALYSIS OF CEMENT MORTAR LIFE IN ANCHOR SYSTEM

3.1   *Experimental design*

Because the water of the underground works in the erosion usually contains Na^+, H^+, Cl^-, SO_4^{2-} plasma, select the soluble Na_2SO_4, HCl, H_2SO_4 as the medium, the preparation of corrosion solution for corrosion testing.

In order to obtain the long time corrosion results in a relatively short period of time, increase the concentration of the medium, but high concentration will affect the authenticity and practicality of the test results. As a result, the concentration of SO_4^{2-} in Na_2SO_4 solution was 1%, 1.5%, 3% respectively; in the acid solution corrosion test, the concentration of HCl was 1%, 3%, 5% respectively and the concentration of H_2SO_4 is 1.0% and 3.0% and 5.0%.

In order to get the relationship between the strength loss of the cement mortar specimen and the corrosion time, the test age of the salt solution corrosion test is 6 months, 12 months, 18 months and 24 months respectively; the age of the acid solution corrosion test is 2 months, 4 months, 6 months, 8 months, 10 months and 12 months respectively.

Standard test piece size of cement mortar is $7.07 \times 7.07 \times 7.07$ cm^3 and the strength grade is M5, using P.O42.5 ordinary portland cement and ordinary sand, by weight with the ratio of cement: sand: water = 1:8.6:1.5, in standard curing condition maintenance. One group is the control group, the standard test method is used to determine the standard value of compression, so as to compare with the strength of the corroded specimens. The prepared cement mortar specimens is put into different solutions, and the mortar samples soaked in solution are taken out according to the predetermined test age, and the compressive strength after corrosion is determined.

3.2 Test result analysis

Compressive strength is an important standard to measure the quality of cement mortar. The compressive strength of the specimens under the corresponding test age is measured, and the strength loss ratio is obtained, which is the standard to measure the corrosion degree of the mortar. The test data are shown in Table 4 and Table 5.

The relationship between strength loss rate of mortar in different concentration Na$_2$SO$_4$ solution and time is shown in figure. Cement mortar test block in the erosion of the salts medium, at the early stage of corrosion, strength loss rate is negative, the compressive strength increases, phenomenon of "strengthening corrosion" is observed, and there is a "corrosion strengthening stage" which is a period from strength beginning to increase to returning to the original strength. With the increase of the concentration of corrosive medium, the corrosion hardening stage becomes shorter, and with the loss of concentration, the corrosion resistance is reduced.

In the acid medium erosion, the strength loss rate and the corrosion time of the mortar test block are linear change, as shown in figure. The medium concentration is bigger, the slope of the line in the figure is bigger, and the loss of strength of cement mortar growth rate is faster. At the same medium concentration, the corrosion damage of sulfuric acid is larger than that of hydrochloric acid.

3.3 Grey prediction of serviceable life of cement mortar in the test environment

The test salt solution of mortar strength loss rate is low, there exists the "corrosion strengthening stage" and the strength loss rate is negative at the beginning, and the largest strength

Table 4. Mortar strength loss rate in salt solution.

Corrosive medium	Concentration/%	Test age/month			
		0.5	1	1.5	2
SO$_4^{2-}$	1.0	−3.1	−3.8	−2.9	−1. 8
	1.5	−4.2	−4.4	−1.9	0.6
	3.0	−5.1	−4.3	0.5	3.2

Table 5. Strength loss ratio of mortar in acid solution.

Corrosive medium	Concentration/%	Test age/month					
		2	4	6	8	10	12
HCl	1.0	2.6	3.5	8.1	7.6	13.6	13.4
	3.0	3.1	8.3	10.4	16.3	17.5	22.5
	5.0	4.7	12.4	14.9	22.2	23.8	32.1
H$_2$SO$_4$	1.0	3	3.9	9.5	9	15	13.5
	3.0	5.1	8.1	17.6	18.3	25.4	25.8
	5.0	9.1	13	24.1	24.6	36.2	41.4

loss rate in the salt solution is just 3.2% at the end of the trial. As a result, it is necessary to do a medium and long term forecast until the strength loss rate of mortar reduced to a threshold value. When –A≤0.3, the GM (1,1) can be used in medium and long-term forecast. In the built model, the –A value is always less than 0.05. Therefore, it is feasible to predict the strength loss rate of mortar in the long term by using GM (1,1) model.

According to the strength grade of the mortar in masonry engineering, the experiment is biased to the safe choice of $0.75\,f_{m,k}$ as the threshold of the strength failure of the mortar. If the average compressive strength of the test specimen is f_m, the threshold value of the strength loss rate of the model can be obtained that is 25%. By using the MATLAB program to calculate, as shown in Table 6, the minimum value of the average precision of the model is 0.9694, and the maximum value of the average level deviation is 0.0532.

As shown in Fig. 9 to Fig. 11, in the Na_2SO_4 solution, the strength loss ratio of mortar in the two years is used as the forecast data sequence. The obtained model values are very close to the known data, and there are even many near coincidence in the Fig. 9. In HCl and H_2SO_4 solution, the strength loss rate of one year is predicted as the forecast data sequence. The obtained model is also very close to the known data, which further proves the reliability of the prediction results.

According to GM (1,1) model and the experimental values of the test results, the corrosion state of the mortar strength loss rate of Na_2SO_4 solution after the test period of 24 months and HCl and H_2SO_4 solution after the test cycle of 12 months was further predicted by the grey prediction, which reached and exceeded the threshold value.

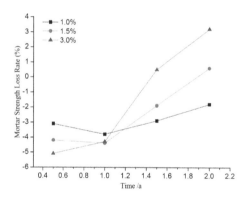

Figure 6. Mortar strength loss ratio in Na_2SO_4 solution.

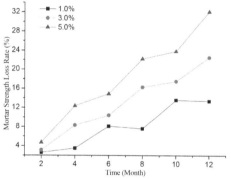

Figure 7. Mortar strength loss ratio in HCl solution.

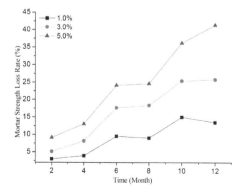

Figure 8. Mortar strength loss ratio in H_2SO_4 solution.

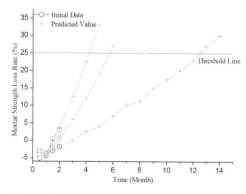

Figure 9. Mortar strength loss ratio in Na_2SO_4 solution.

162

Table 6. Parameters and test index of cement mortar strength loss rate of GM (1,1) model.

Corrosive medium	Concentration /%	Raw data sequence 1	2	3	4	5	6	Shift value	A	B	Average precision	Average level deviation
SO_4^{2-}	1.0	-3.1	-3.8	-2.9	-1.8	/	/	50	-0.0212	44.6848	0.991	0.014
	1.5	-4.2	-4.4	-1.9	0.6	/	/	50	-0.0520	42.0575	0.995	0.020
	3.0	-5.1	-4.3	0.5	3.2	/	/	50	-0.0747	41.0480	0.989	0.036
HCl	1.0	2.6	3.5	8.1	7.6	13.6	13.4	50	-0.0429	50.8244	0.978	0.043
	3.0	3.1	8.3	10.4	16.3	17.5	22.5	50	-0.0564	53.0870	0.988	0.029
	5.0	4.7	12.4	14.9	22.2	23.8	32.1	50	-0.0684	55.8418	0.982	0.043
H_2SO_4	1.0	3.0	3.9	9.5	9.0	15	13.5	50	-0.0425	51.9466	0.972	0.052
	3.0	5.1	8.1	17.6	18.3	25.4	25.8	50	-0.0614	55.5814	0.969	0.048
	5.0	9.1	13.0	24.1	24.6	36.2	41.4	50	-0.0882	56.6979	0.973	0.053

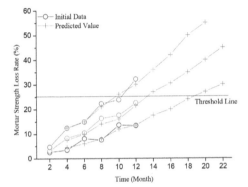

Figure 10. Mortar strength loss ratio in HCl solution.

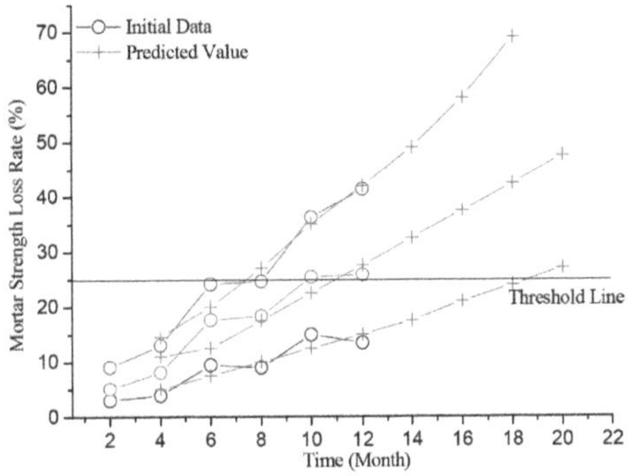

Figure 11. Mortar strength loss ratio in H_2SO_4 solution.

The results show that the development of the prediction curve shape is gradual, natural and harmonious, and there is no distortion. According to the basic principle of grey prediction, the predicted results after the test cycle have the same confidence with those in the test period.

From the table, In terms of different corrosive media, of service life of mortar under test conditions, the longest is in SO_4^{2-} ion (When the concentration is 1%, the service life is 12.4a), the shorter is in HCl, and the shortest is in H_2SO_4 (When theconcentration is 5%, the service life is only half a year's). Generally speaking, the service life is not long. However, what is done here is Accelerated corrosion test, does not involve the specific corrosion environment of the project.

Based on the service life of cement mortar specimens under test conditions, using GM (1,1) model to calculate the relation curve and function expression between the service life and the ion concentration, And then substituting the corresponding parameters of the concrete engineering in corrosive environment into function expression, we are able to make more reliable prediction of the service life of the anchor structure in the project. The function expression for service life and ion concentration is:

$$y = A + Be^{Cs}$$

where y = service life, a; s = corrosion solution concentration, %; A, B, C = coefficient associated with corrosion environment.

163

4 CONCLUSIONS

In this paper, through the method combining laboratory experiments and theoretical analysis, using grey system theory of laboratory test results for the further analysis and research, to analyze and study comprehensively the durability and service life of anchor bolt, draw the following conclusions:

1. In four different corrosion environment of closed wet, permanent immersion, alternating wet and dry, weak acid aqueous solution, the weight loss rate of anchor bolt body in weak acid aqueous solution is the largest, and that in the closed moist environment is the smallest. In no matter what kind of test environment, the weight loss rate of anchor is increased with time, but the increase speed is decreased with time.
2. Under the salt medium erosion, the compressive strength of cement mortar test block is increased at the initial stage of corrosion. The strength loss is accelerated with the increase of the concentration of corrosive medium, and the corrosion resistance of the concrete block is decreased.
3. In the acid medium erosion, the strength loss rate and the corrosion time of the mortar test block are linear change. The bigger medium concentration is, the faster loss of strength of cement mortar growth rate is. At the same medium concentration, the corrosion damage of sulfuric acid is larger than that of hydrochloric acid.
4. The service life of mortar under accelerated corrosion tests, the longest is in SO_4^{2-} ion, the shorter is in HCl, and the shortest is in H_2SO_4. In different concentrations, the greater the concentration is, the shorter the life is.

REFERENCES

Ballim Y., Reid J.C., Kemp A.R. Deflection modeling for RC beams under simultaneous load anf steel corrosion [J].Cement and Concrete Research, 2001, 53(3):171–181.
Capozucca R. Identification of damage in reinforced concrete beams subjected to corrosion [J]. ACI Structral Journal, 2000, 97(6):902–909.
Cheng Liangkui. Present status and development of ground anchorages [J]. China Civil Engineering Journal, 2001, 34(3):7–12.
Williamson S.J., Clark L.A. Pressure required to cause cover cracking of concrete due to reinforcement corrosion[J].Cement and Concrete Research, 2000, 52(6):455–467.
Zeng Xianming, Chen Zhaoyuan, Wang jingtao. Research on safety and durability of bolt and cable-supported structures [J].Chinese Journal of Rock Mechanics and Engineering, 2004, 23(13):2235–2242.

Hydraulic Engineering IV – Xie (Ed.)
© 2016 Taylor & Francis Group, London, ISBN 978-1-138-02948-4

Simulation of transport of sediment and heavy metal in a bay system

Y. Yan & L. Lei
China Waterborne Transport Research Institute, Beijing, China

Y.J. Jin
Huadian Electric Power Research Institute, Hangzhou, China

ABSTRACT: This paper applied a dynamic model to simulate transport of sediment and four trace metals (cadmium, copper, lead, and zinc) in Deep Bay. The results indicated that the whole bay was under sedimentation, and most heavy metals from river inflow mainly deposited in river mouth. However, metals were unevenly distributed in neighbored areas and heavily accumulated in some sensitive areas. The results could be used for interpreting why the ecological degradation in Mai Po Marshes was worse than its adjacent area, Futian National Nature Reserve.

1 INTRODUCTION

Deep Bay is situated between Shenzhen (a special economic zone of China) and Hong Kong (Figure 1). Futian National Nature Reserve Site (FTRS) and Mai Po Marshes and Inner Deep Bay Ramsar Site (MPRS) are located in the southeast of Deep Bay. FTRS and MPRS serve as refueling sites for over 60,000 migratory birds in every spring and autumn (Man et al., 2004).

Figure 1. General localization of the Deep Bay and the sampling stations (SZR, DSR, YLC, and TSW indicate Shenzhen River, Dasha River, Yuen Long Creek, and Tin Shui Wai Creek).

With rapid economic developments, polluted water and contaminative sediments were continuously introduced into Deep Bay. The heavy-metal accumulation led to degradation of species diversity in Deep Bay (Cheung et al., 2003). However, it was very interesting that neighbored areas (just separated by Shenzhen River) had quite different responses to the similar human interferences (Lei et al. 2013). The degradation in MPRS was worse than that in its adjacent areas, FTRS (Cai et al., 2003). No literature has interpreted the interesting phenomenon. With this in mind, a dynamics model, Environmental Fluid Dynamics Code (EFDC) model, was applied to interpret the phenomenon in this paper.

2 METHODS

2.1 Governing equation

Hydrodynamics models, containing momentum and continuity equations, are constructed on the basis of hydrostatic approximation and Boussinesq approximation. Equations are formulated in curvilinear and orthogonal coordinates and sigma coordinate system in the horizontal and vertical, respectively. Since heavy metals have little affinity for coarse particles, cohesive particle is considered in this paper. More details about the hydrodynamics, sediment, and heavy-metal transport governing equations can be found in Ji et al. (2001) and Ji (2008).

2.2 Model configuration

Flow is tidal driven in Deep Bay. Tidal elevation at Chiwan station is set as open boundary condition. Freshwater inputs include discharges from Shenzhen River, Dasha River, Yuen Long Creek, and Tin Shui Wai Creek. Table 1 lists the average discharges entering Deep Bay from these rivers during the wet and dry seasons. Deep Bay was meshed by a grid with 2512 horizontal curvilinear cells.

3 RESULTS AND DISCUSSION

3.1 Hydrodynamics

Figure 2 shows comparison of simulated and observed tidal elevation at station DJT. Differences between model results and field data are less than 0.441 m. The computed tidal elevations are satisfactory. Figure 3 presents comparison of calculated and measured tidal velocity at station 1. Generally, the model results agree well with the observed tidal velocity.

Table 1. Flow discharge, Suspended Sediment (SS), and heavy-metal concentrations at river inlets to Deep Bay.

Parameter	Shenzhen River	Dasha River	Yuen Long Creek	Tin Shui Wai Creek
Q_{wet} (m³/s)	13.9	4.67	2.46	1.64
Q_{dry} (m³/s)	1.4	0.47	0.34	0.23
SS_{wet} (g/L)	50	40	35	30
SS_{dry} (g/L)	30	10	10	10
Cd (µg/L)	0.5	0.2	0.25	0.25
Cu (µg/L)	15	12.2	32.8	12
Pb (µg/L)	5.4	3.2	7.7	8
Zn (µg/L)	50	76	140	40

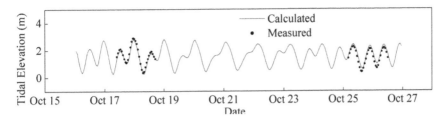

Figure 2. Comparison of calculated and measured tidal elevations at stations DJT.

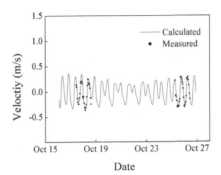

Figure 3. Comparison of calculated and measured velocity at station 1#.

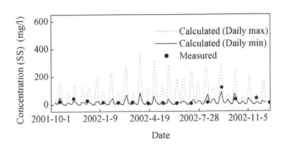

Figure 4. Comparison of predicted and measured Suspended Sediment (SS) concentrations at station 2.

3.2 Cohesive sediment transport

Figure 4 shows comparison of modeled and monitored suspended sediment concentrations at station 2. The model reproduces main features of sediment transport in Deep Bay.

3.3 Heavy-metal transport

In Figure 5, comparison of simulated and measured concentration of dissolved cadmium, copper, lead, and zinc at station 3 is indicated. EFDC reasonably reproduces general trend of observed values.

Figure 6 shows heavy-metal accumulation rate in Deep Bay. It is indicated that the accumulation rates of Cd, Cu, Pb, and Zn on the bed of Mai Po Marshes are, respectively, 2.7, 3.3, 3.4, and 2.8 times higher than for Futian National Natural Reserve. The topography causes this phenomenon. When tide rises, Tsim Bei Tsui reduces current speed at Mai Po

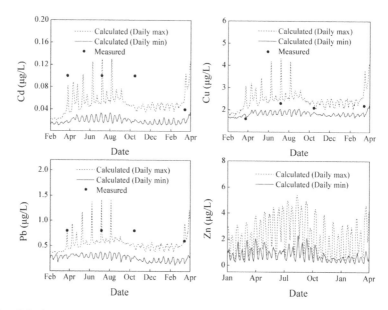

Figure 5. Calculated and measured concentrations of dissolved heavy metal at station 3 from 2002 to 2003.

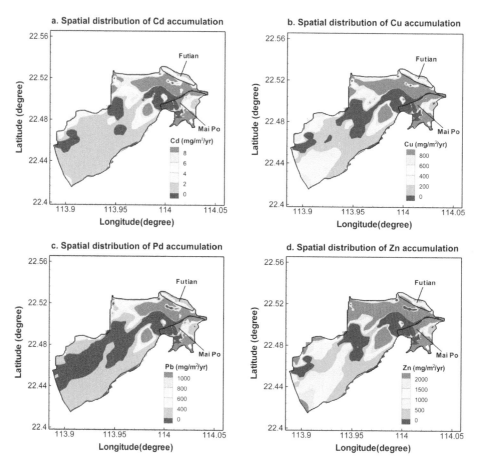

Figure 6. Spatial distribution of heavy-metal accumulation on the bed of Deep Bay.

Marshes, whereas FTRS is scoured without any shield against tide. Lower current speeds at Mai Po Marshes lead to higher sedimentation and heavy-metal accumulation rates.

It was reported that heavy metals damaged macro-invertebrate populations in Mai Po Marshes (Laboratory of Environmental Toxicology of the University of Hong Kong, 2003). The heavy-metal accumulation was the main factor leading to macro-invertebrate species decrease. Hence, EFDC simulation of heavy-metal accumulation helps to provide a credible reason for the different degradation of species diversity between the Shenzhen and Hong Kong sites. This is consistent with the report of Cai et al. (2003).

4 CONCLUSION

The EFDC model was used to simulate heavy-metal transport in Deep Bay. Rising and ebbing tide processes, suspended solid and heavy-metal dispersions were successfully reproduced. Then, calibrated model revealed that heavy-metal accumulation rate at Mai Po Marshes was three times more than that in FTRS. Hydrodynamic characteristics led to this phenomenon. Furthermore, the significant difference in metal accumulation led to different species diversity degradation between Shenzhen and Hong Kong sites.

REFERENCES

Cai, L.Z., Tam, N., Wong, T., Ma, L., Gao, Y. & Wong, Y.S. 2003. Using benthic macrofauna to assess environmental quality of four intertidal mudflats in Hong Kong and Shenzhen Coast. *ACTA Oceanologica Sinica* 22(2), 309–319.

Cheung, K.C., Poon, B., Lan, C.Y. & Wong, M.H. 2003. Assessment of metal and nutrient concentrations in river water and sediment collected from the cities in the Pearl River Delta, South China. *Chemosphere* 52(9), 1431–1440.

Ji, Z.G. 2008. Hydrodynamics and Water Quality: Modeling Rivers, Lakes, and Estuaries. Hoboken: John Wiley & Sons.

Ji, Z.G., Morton, M.R. & Hamrick, J.M. 2001. Wetting and drying simulation of estuarine processes. *Estuarine Coastal and Shelf Science* 53(5), 683–700.

Laboratory of Environmental Toxicology of the University Of Hong Kong, 2003. *Final report: baseline ecological monitoring programme for the Mai Po and Inner Deep Bay Ramsar Site.* HongKong: University Of Hong Kong.

Lei, L., Sun, J.S., Borthwick, A.G.L., Fang, Y., Ni, J.R. 2013. Dynamic Evaluation of Intertidal Wetland Sediment Quality in a Bay System. *Journal of Environmental Informatics* 21(1):12–22

Man, K.W., Zheng, J.S., Leung, A., Lam, P., Lam, M. & Yen, Y.F. 2004. Distribution and behavior of trace metals in the sediment and porewater of a tropical coastal wetland. *Science of The Total Environment* 327(1–3), 295–314.

Hydraulic Engineering IV – Xie (Ed.)
© 2016 Taylor & Francis Group, London, ISBN 978-1-138-02948-4

The effect of an Enhanced Ecological Floating Bed (EEFB) on phytoplankton community in an urban tidal river

Hong Pan
Institute of Hydrobiology, Jinan University, Guangzhou, China
School of Public Health, Zunyi Medical College, Zunyi, China

Yang Yang
Institute of Hydrobiology, Jinan University, Guangzhou, China

ABSTRACT: Phytoplankton samples were collected monthly from May to October in 2009 from the Demonstration Plots (DP) and Control Plots (CP) of an Enhanced Ecological Floating Bed (EEFB), which was set up in three segments of Luocun River. A total of 171 phytoplankton species were identified in the river. Out of them, 145 and 135 species were detected in the DP and CP, respectively. The richness, abundance, biovolume, Margalef index (D_M) and Shannon-Wiener index (H') were higher in the DP than those in the CP in 94.44%, 66.67%, 66.67%, 94.44% and 88.89% of the paired-samples, respectively. The similarity coefficient ranged from 0.25 to 0.78 between the CP and DP in three segments. The results of a paired-samples T test showed that the richness and D_M were significantly different ($p < 0.05$) between the DP and CP in three segments. The results of the relative dominance (D_r), which indicates a change in dominant species, showed that the community has altered in 50% of the paired-samples. The results indicated that the EEFB could improve the structure and stability of the phytoplankton community.

1 INTRODUCTION

The restoration of many rivers in lowland areas aims at re-creating morphological features and increasing the habitat diversity. However, it is rarely feasible for a complete restoration, because the lowland rivers are usually located in areas being used for agricultural production or urban settlements (Pedersen, Friberg, et al., 2007). It was very difficult to reconstruct river morphology in the coastal areas of South China, where the agricultural production and urban settlements are located along the riversides.

In relative static water (lakes, ponds and reservoirs), phytoplankton community acquires nutrition and intercepts suspended solids directly from the water column with vegetation of the floating islands or beds. It can change the local habitat in which their roots are suspended (Headley and Tanner, 2006). Shading with floating islands/beds (O'Farrell, Pinto, et al., 2009) and competition with other vegetation for nutrients and light can reduce algal blooms (Lee and Kwon, 2004, Nakamura and Mueller, 2008). Therefore, the floating islands/beds can be used as a restoration tool for improving the environment of relative static water.

Some studies have focused on reintroduction of keystone species such as fish, plants and macrozoobenthos for ecological restoration of rivers (Neumann, 2002, Nienhuis, Bakker, et al., 2002, Raat, 2001). However, it was difficult to implement in most areas of China (Ni and Liu, 2006). The phytoplankton, as primary producers, support food webs and play a central role in carbon, nutrient and oxygen cycling (Paerl, Rossignol, et al., 2010). Therefore, a change in phytoplankton community structure could adversely affect the ecosystem stability. However, the variation in phytoplankton community may be different in running water (such as rivers and streams) after establishing an Enhanced Ecological Floating Bed (EEFB). In this study, we installed approximately 3108 m² of the EEFB in an urban river of Foshan city, Guangdong province, China. Objectives of this study were to compare the value of richness, abundance, biovolume, relative dominance and diversity indices between demonstration and control plots (DP and CP, respectively) to evaluate beneficial effects of EEFB for the restoration of phytoplankton community.

2 METHODS

2.1 Study area

Luocun River (LCR) is a tidal channel. The EEFB were installed at three segments of the urban river, which are Luocun (LC), Liangan (LA) and Wangzhi (WZ) segments (Fig. 1). The banks of these segments are vertically solidified and straightly channelized by the concrete revetment for enhancing the ability of flood control. The color of water in these segments was black or gray. The segments were releasing strong bad smell due to receiving a large amount of industrial wastes and urban sewages in recent years.

2.2 Design and installation of EEFB

A rectangular unit of EEFB (approximately 4.0 m × 1.75 m) was composed of 60 polyethylene (HDPE) quadrate modules (0.33 m × 0.33 m). Every module had a hole in the center for planting basket. *Cyperus alterniflius, Pontederia cordata, Canna generalis, Thalia dealbata, Iris pseudoacorus, Iris tectorum* and *Arundo donax* var. *versicolor* were planted and filled with haydite or gravel in the basket. Total installation of EEFB encompasses approximately 3108 m² in the urban river. The size of the EEFB in LC, LA and WZ segments was 2 128 m², 560 m² and 420 m², respectively.

2.3 Sampling and analyses

Some demonstration plots (DP) and Control Plots (CP) were selected at the inflow and outflow of the EEFB, and CP and DP were installed at the straight part of the segment. Samples were collected at 0.5 m below the water surface at sites of the CP and DP from May to October, 2009. Phytoplankton samples were preserved with a Lugol solution at a final concentration of 1% (v/v). The species in the water samples were identified, when possible,

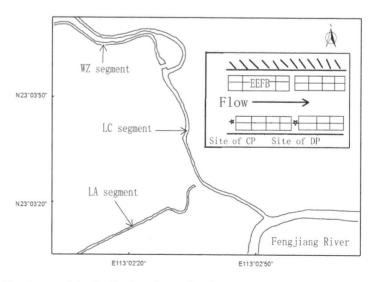

Figure 1.   Sketch map of the distribution of sampling sites and segments in the Luocun River.

based on published descriptions (Akiyama, Ioriya, et al., 1981, Hu and Wei, 2006, John, Whitton, et al., 2002, Wehr and Sheath, 2003). Counting was performed with an Olympus CX31 microscope as previously described. Biovolume was calculated as previously described (Olenina, Hajdu, et al., 2006). The dominant species was determined with the relative dominance (D_r) in the community. It was calculated by the formula: $D_r = \dfrac{(n_i/N)(b_i/B)}{\sum (n_i/N)(b_i/B)}$, where the (n_i/N) and (b_i/B) are the relative abundance and biovolume of ith species, respectively.

The Margalef index (D_M) and Shannon-Wiener index (H) (Wilhm, 1968) were calculated by the formula: $D_M = \dfrac{S-1}{\ln N}$ and $H = -\sum_{i=1}^{S} (\dfrac{n_i}{N}) \ln(\dfrac{n_i}{N})$, respectively. Where the S, N and n_i are the species richness, total abundance and ith species abundance, respectively. The similarity coefficient (C_S) was calculated by the formula: $C_S = \dfrac{2a}{b+c}$, where a, b and c was the number of common species, richness of the CP and DP, respectively.

The data were calculated by Microsoft Office Excel 2003. The paired-samples T test was used to determine the difference in richness, abundance, biovolume, D_M and H between the CP and DP. The significance level was $p < 0.05$. Statistical analyses were performed with SPSS 18.0 (SPSS Inc.).

3   RESULTS

3.1   *Components of the phytoplankton*

There were 171 species recorded in the LCR, which belong to 5 phylum, 8 classes, 16 orders, 33 families and 73 genera. Among them, 145 and 135 species were detected in the DP and CP, respectively. The richness was higher in the DP than that in the CP in 94.44% of the paired-samples (Fig. 2). The paired-samples T test results showed that richness was significantly different ($p < 0.05$) between the DP and CP (Table 1). Some genera such as *Surirella*, *Synedra*, *Hydrococcus*, *Tetrastrum* and *Staurastrum* were detected only in the DP samples. Some other genera such as *Microcystis*, *Pyrobotrys* and *Strombomonas* were detected only in the CP samples.

The range of abundance was found to be 1.71–83.71×10^6 cell.L^{-1}, 1.79–69.65×10^5 cell.L^{-1} and 0.06–734.1×10^5 cell.L^{-1} in LA, LC and WZ segments, respectively. The abundance

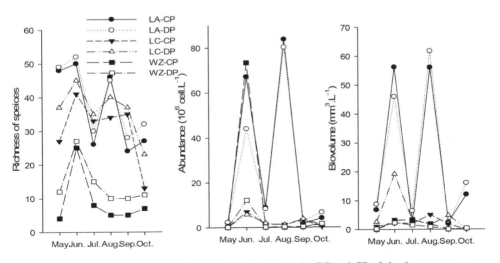

Figure 2.   Richness of species, abundance and biovolume in the DP and CP of the three segments.

173

was higher in the DP than that in the CP in 66.67% of the paired-samples (Fig. 2). The paired-samples T test results showed that the abundance was not significantly different ($p > 0.05$) between the DP and CP (Table 1).The range of biovolume was 1.87–61.75 mm³.L⁻¹, 0.10–19.21 mm³.L⁻¹ and 0.01–3.30 mm³.L⁻¹ in LA, LC and WZ segments, respectively. The biovolume was higher in the DP than that in the CP in 66.67% of the paired-samples (Fig. 2). The paired-samples T test results showed that the biovolume was not significantly different ($p > 0.05$) between the DP and CP (Table 1).

Euglena mutabili, Lepocinclis texta, L. steinii, L. ovum, Chlorella vulgaris, Cryptomonas eros, Fragilaria brevistriata, Ankistrodesmus aciculars and *Merismopedia punctata* were the dominant species with the highest D_r in the LCR. The D_r of the dominant species apparently changed in 66.7% of the paired-samples while flowing out of the EEFB. Especially, the community structure has altered in 50% of the paired-samples (Table 2).

3.2 Diversity indices

The range of D_M was 1.56–3.28, 0.95–2.86 and 0.33–1.59 in LA, LC and WZ segments, respectively. The D_M was higher in the DP than that in the CP in 94.44% of the paired-samples (Fig. 3). The paired-samples T test results showed that the D_M was significantly different ($p < 0.05$) between the DP and CP in the three segments (Table 1). The range of H was 1.93–4.60, 1.86–4.66 and 0.20–3.46 in LA, LC and WZ segment, respectively. The H was higher in the DP than that in the CP in 88.89% of the paired-samples (Fig. 3). The

Table 1. The p value of paired-samples T test between the DP and CP (2-tailed).

	LA segment[a]	LC segment[a]	WZ segment[a]	LC River[b]
Richness	0.042	0.013	0.002	0.000
Abundance	0.370	0.867	0.368	0.211
Biovolume	0.751	0.308	0.132	0.359
D_M	0.034	0.017	0.002	0.000
H	0.099	0.113	0.005	0.000

[a]df = 5; [b]df = 17.

Table 2. The D_r of dominant species in the CP and DP.

Sampling month	Dominant species	LA segment	LC segment	WZ segment
May	*Euglena mutabilis*	0.40/0.22		
	Lepocinclis steinii			0.60/0.54
	Lepocinclis texta		0.76/0.73	
Jun.	*Chlorella vulgaris*	0.52/0.72	0.60/0.11	
	Euglena mutabilis		⁻[a]/0.64	
	Merismopedia punctata			0.99/0.25
	Ankistrodesmus aciculars			0.00 [b] /0.29
Jul.	*Chlorella vulgari*	0.99/0.17	0.23/0.19	
	Euglena mutabili	0.00 /0.45	0.22/0.60	0.95/0.68
Aug.	*Lepocinclis ovum*			0.05/0.38
	Euglena mutabilis	0.46/0.42	0.57/0.03	0.95/0.33
	Cryptomonas erosa		0.22/0.37	
Sep.	*Euglena mutabilis*	0.37/–		
	Chlorella vulgaris	0.24/0.49	0.55/0.41	0.77/0.71
Oct.	*Chlorella vulgaris*	0.16/0.43		0.95/0.89
	Euglena mutabilis	0.77/0.41	0.32/0.01	
	Fragilaria brevistriata		0.16/0.74	

[a]undetected in the samples, [b]the value is too low.

paired-samples T test results showed that the H was significantly different ($p < 0.05$) between the DP and CP only in the WZ segment (Table 1). The range of C_S was 0.61–0.71, 0.56–0.71 and 0.25–0.78 in LA, LC and WZ segments, respectively. The mean value was 0.64, 0.64 and 0.57 in LA, LC and WZ segments, respectively (Fig. 3).

4 DISCUSSION

Competition, bio-degradation, nutrition constraints, allelopathy and shading are generally the main mechanisms to reduce algal blooms in relative static water by floating islands/beds (Hoeger, 1988, Lee and Kwon, 2004, Li, Song, et al., 2010, Nakai, Inoue, et al., 2001, Nakai, Zou, et al., 2008, Nakamura and Mueller, 2008, Nakamura and Shimatani, 1997, O'Farrell, Pinto, et al., 2009, Søndergaard and Moss, 1997). However, in this study, the abundance and biovolume were not significantly different ($p > 0.05$) between the DP and CP (Table 1). The reasons could be as follow: (I) The competition of nutrient was almost inexistent between phytoplankton and other vegetation in urban rivers with higher concentration of nutrients (Yang, Yang, et al., 2011). (II) The competition for light and shading was almost inexistent in the running water. (III) The effects of allelopathy and degradation by algae-lysing bacteria could have been ignored due to the turbulence. In contrast, interruptions by haydite or gravel in the plant basket could have created more minuscular fluvial dead-zones in the EEFB. It is reported that obtaining nutrition and intercepting suspended solids with vegetation roots could improve water quality and transparency in the microhabitat (Headley and Tanner, 2006, Yang, Yang, et al., 2011). The phytoplankton, which has fast growth rates, can rapidly respond to the habitat perturbations (Paerl, Rossignol, et al., 2010). Therefore, the abundance/biovolume of some species had increased, which were adapting to the new habitats. On the other hand, those incompatible with the changing environment could have decreased. At the higher velocity condition, the species were taken out from the EEFB due to river flow. Therefore, the abundance was significantly higher in the DP than that in the CP in some paired-samples (e.g. at the LC and WZ segments in June).

In this study, species richness was significantly ($p < 0.05$) higher in the DP (Fig. 2) compared to that in the CP (Table 1). In addition, some genera such as *Surirella*, *Synedra*, *Hydrococcus*, *Tetrastrum* and *Staurastrum* were detected only in the DP, and some others such as

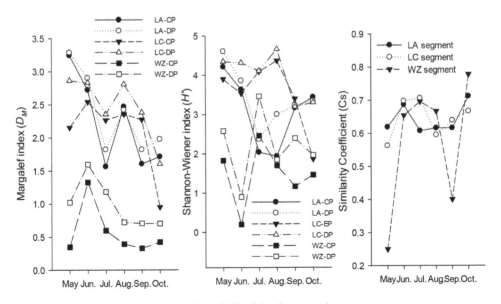

Figure 3.   The D_M, H and C_S in the DP and CP of the three segments.

175

Microcystis, *Pyrobotrys* and *Strombomonas*) only in the CP. Therefore, the EEFB is not only beneficial for increasing richness but also for changing the composition of phytoplankton community in the urban river.

The dominant species can effectively adapt to the changing habitats, capable of resisting disturbances and keep community structure stable (Guo, Ma, et al., 2010). In our study, the results of D_r showed that the phytoplankton community structure had 3 types of alteration (Table 2): mono-dominant to other mono-dominant (e.g. the samples of July and September in the LA segment), mono-dominant to multi-dominant (e.g. the samples of June and August in the WZ segment) and multi-dominant to mono-dominant (e.g. the samples of July in the LC segment). It indicated that the EEFB could change the microhabitats and stability of the community structure of phytoplankton in the urban river. However, as a tidal channel, the daily change in flow velocity of the river was very huge in the LCR. Except for the interception by the EEFB, other effects such as competition, allelopathy and shading for phytoplankton could have been neglected when the samples were collected at the faster flow velocity. Therefore, the dominant species apparently changed in 33.3% of the paired-samples (e.g. the samples of May in the LC segment), indicating that the change in phytoplankton community might have been significantly influenced by the flow velocity.

The D_M and H' were higher in the DP than those in the CP in 94.44% and 88.89% of the paired-samples, respectively. It indicates that the change in richness and abundance by the EEFB could have increased the phytoplankton diversity. The results of C_S showed that the range was from 0.25 to 0.78, indicating that the change in species replacement was significant along the environment gradient, which was created by the EEFB.

In addition, Yang et al. (2011) chose one of the segments to study the zooplankton community. Their study showed that the richness, abundance and diversity of zooplankton in the DP were significantly higher than those in the CP most of the time. Therefore, we consider that the grazing with zooplankton could be one of the indirect factors for the change in phytoplankton community by the EEFB.

The ecological status has to be stated mainly on the basis of the biological quality elements which are species composition, abundance and biomass of the phytoplankton (Wasmund, Göbel, et al., 2008). The diversity, species composition, and the pattern of interactions among species are the three basic factors in deciding the stability of ecological communities (Ives, Dennis, et al., 2003). Therefore, based on the results, we consider that the EEFB could improve the structure and stability of the phytoplankton community.

ACKNOWLEDGMENTS

We are grateful to the members of our team for their assistance in collecting the samples. This research was supported by "Major science and technology projects of water pollution control and governance of China (2008ZX07211–003)".

REFERENCES

Akiyama, M., T. Ioriya, K. Imahori, H. Kasaki, S. Kumano, H. Kobayasi, et al., editor. 1981. Illustrations of the Japanese Freshwater Algae. Uchidarokakuho Publishing Co. Ltd., Tokyo.

ASCE, Ahupua'a. 1997. Water purification and environmental enhancement by the floating wetland. Proceeding of the Asia Waterqual'97 in Korea, Seoul Korea.

Guo, Q., K. Ma, L. Yang, Q. Cai and K. He. 2010. A comparative study of the impact of species composition on a freshwater phytoplankton community using two contrasting biotic indices. Ecological Indicators 10: 296–302.

Headley, T.R. and C.C. Tanner. 2006. Application of floating wetlands for enhanced stormwater treatment: a review. NIWA Client Report prepared for the Auckland Regional Council.

Hoeger, S. 1988. Schwimmkampen: Germany's artificial floating islands. Journal of Soil and Water conservation 43: 304–306.

Hu, H.-J. and Y.-X. Wei, editor. 2006. The Freshwater Algae of China: Systematics, Taxonomy and Ecology. Science Press, Beijing.

Ives, A.R., B. Dennis, K.L. Cottingham and S.R. Carpenter. 2003. Estimating community stability and ecological interactions from time-seriesdata. Ecological Monographs 73(2): 301–330.

John, D.M., B.A. Whitton and A.J. Brook. 2002. The Freshwater Algal Flora of the British Isles Cambridge University Press, Cambridge.

Lee, E.J. and O.B. Kwon. 2004. The effects of floating islands planted with various hydrophytes for water quality. Rep. Res. Edu. Ctr. Inlandwat. Environ (2): 121–126.

Li, X.-N., H.-L. Song, W. Li, X.-W. Lu and O. Nishimura. 2010. An integrated ecological floating-bed employing plant, freshwater clam and biofilm carrier for purification of eutrophic water. Ecological Engineering 36: 382–390.

Lund, J.W.G., C. Kipling and E.D.L. Cren. 1958. The inverted microscope method of estimating algal numbers and the statistical basis of estimations by counting. Hydrobiologia 11: 143–170.

Nakai, S., G. Zou, X. Song, Q. Pan, S. Zhou and M. Hosomi. 2008. Release of anti-cyanobacterial allelochemicals from aquatic and terrestrial plants applicalbe for artificial floating islands. Journal of Water and Environment Technology 6: 55–63.

Nakai, S., Y. Inoue and M. Hosomi. 2001. Algal growth inhibition effects and inducement modes by plant-produced phenols. Water Research 35: 1855–1859.

Nakamura, K. and G. Mueller. 2008. Review of the performance of the artificial floating island as a restoration tool for aquatic environment. World Environmental and Water Resources Congress.

Neumann, D. 2002. Ecological rehabilitation of a degraded large river system-considerations based on case studies of macrozoobenthos and fish in the lower Rhine and Its catch area. Internat. Rev. Hydrobiol. 87: 139–150.

Ni, J.-r. and Y.-y. Liu. 2006. Ecological rehabilitation of damaged river system. Shuili Xuebao 37(9): 1029–1037, 1043.

Nienhuis, P.H., J.P. Bakker, A.P. Grootjans, R.D. Gulati and V.N.D. Jonge. 2002. The state of the art of aquatic and semi-aquatic ecological restoration projects in the Netherlands. Hydrobiologia 478: 219–233.

Olenina, I., S. Hajdu, L. Edler, A. Andersson, N. Wasmund, S. Busch, et al. 2006. Biovolumes and size-classes of phytoplankton in the Baltic Sea. HELCOM Baltic Sea Environment Proceedings 106: 1–21.

O'Farrell, I., P.D.T. Pinto, P.L. Rodríguez, G. Chaparro and H.N. Pizarro. 2009. Experimental evidence of the dynamic effect of free-floating plants on phytoplankton ecology. Freshwater Biology 54: 363–375.

Paerl, H.W., K.L. Rossignol, S.N. Hall, B.L. Peierls and M.S. Wetz. 2010. Phytoplankton community Indicators of short- and long-term ecological change in the anthropogenically and climatically impacted Neuse River estuary, North Carolina, USA Estuaries and Coasts 33: 485–497.

Pedersen, M.L., N. Friberg, J. Skriver, A. Baattrup-Pedersen and S.E. Larsen. 2007. Restoration of Skjern River and its valley—Short-term effects on river habitats, macrophytes and macroinvertebrates. Ecological Engineering 30: 145–156.

Raat, A.J.P. 2001. Ecological rehabilitation of the Dutch part of the River Rhine with special attention to the fish. Regulated rivers: Research & Management 17: 131–144.

Søndergaard, M. and B. Moss. 1997. Impact of Submerged Macrophytes on Phytoplankton in Shallow Freshwater LakesSpringer.

Wasmund, N., J. Göbel and Bodov.Bodungen. 2008. 100-years-changes in the phytoplankton community of Kiel Bight (Baltic Sea). Journal of Marine Systems 73: 300–322.

Wehr, J.D. and R.G. Sheath. 2003. Freshwater Algae of North America: Ecology and ClassificationAcademic Press, New York.

Wilhm, J.L. 1968. Use of biomass units in Shannon's formula. Ecology 49(1): 153–156.

Yang, F., Y. Yang, H. Pan, D. A, L. Li, Y. Qiao, et al. 2011. Effect of an enhanced ecological floating bed (EEFB) on zooplanton community in a polluted river. Journal of Lake Sciences 23: 498–504.

Hydraulic Engineering IV – Xie (Ed.)
© 2016 Taylor & Francis Group, London, ISBN 978-1-138-02948-4

Dynamic safety management and supervision of activities at public assembly occupancy based on risk assessment

Yunhua Gong

School of Mechanics, Storage and Transportation Engineering, China University of Petroleum, Beijing, China

ABSTRACT: Accidents occurring in assembly occupancies may cause severe consequences which are mainly a large number of casualties, property losses as well as social chaos. To make sure that risks at assembly occupancies are controlled, and risk-control resources are reasonably allocated for each individual activity, safety management and supervision should be conducted dynamically. In other words, safety management and supervision should be adjusted according to the risk level and hazard characters of each activity at assembly occupancies. A dynamic safety management and supervision method based on risk assessment is provided in this paper. Risk assessment on individual activity was conducted by measuring a list of indicators that are categorized into four categories: indicators relating to people, equipment, environment, and safety management. Fuzzy comprehensive evaluation was adopted to get comprehensive risk for each activity. This dynamic safety management and supervision method was used in safety management and supervision plan for a graduation party at a university and was tested to be useful.

1   INTRODUCTION

There are increasing large-scale activities involving a large number of people. And some fatal accidents occurred in recent years. There was a stampede resulting in 36 deaths on 31 December 2014 in Shanghai, China. There are some regulations on how to manage the risk of activities that involve a large number of people in China, such as regulations for the safety management of large-scale mass activities. However, they only provide the common requirements for large-scale activities. That is to say, the safety management and supervision for each activity are nearly the same without considering the differences of risks among the various activities. In this case, the risk control might be inadequate for some high-risk activities, whereas some risk-control measures may be unnecessary for some low-risk activities, which lead to a waste of control resources. So, safety management and supervision plan will be better if it is conducted according to the risk assessment result of each activity at the assembly occupancy.

2   METHOD

2.1   *The characters of assembly occupancies and their hazards*

First, assembly occupancies are characterized by large allocating density of people, which causes difficulty in evacuation (Tong, 2013). People may be differently aged, making evacuation hard. Another character of assembly occupancy is its high risk. Most assembling places are shopping malls, squares and others. To pursue good activity effects, these places are mostly decorated with a large number of flammable decorating materials, colorful lights and a lot of wires carrying electricity. The third character of assembly occupancy is the domino effect of accidents. For example, most of the time, a fire can be followed by a stampede. The consequence of a small accident may be expanded dramatically.

Hazards in assembly occupancies fall into the following categories:

1. Meteorological disaster: Many types of meteorological disasters, including drought, flood, heat waves, winds, tropical cyclones, storm surge disasters, lightning, fog, snow, dust and so on may cause severe consequences when people are crowded in a specific area. On 12 March 1988, there was a football game in Nepal's capital, Kathmandu. Suddenly, the weather changed and big winds along with hail started blowing. So, people fled the stands, and the scene was very chaotic. The official death toll was 73.
2. Infectious diseases: The high concentration of people in crowded places and slow people flow are the perfect environment for the spread of infectious diseases.
3. Stampede: Stampede is a high-possibility accident which may occur in crowded places. There have been a lot of people who have been trampled and died or injured in stampedes. Someone may deliberately create confusion, or the occurrence of earthquakes, fires and other unexpected factors may also cause chaos. Some crowded places have very low lighting and noisy background music, which increase the risk of stampede.
4. Fire: The electricity wires and the devices for making gorgeous effects on the stage have high possibility to cause fire. And, once there is a fire, the consequences will be severe because of the high density of people and difficulty to make an escape.
5. Explosion: Fireworks and smoke effects are often used in large activities. These substances have high potential to explode.
6. Infrastructure failure: Failure or damage to water supply, communication, transportation, gas supply and power supply system may result in the sudden interruption of activities in crowded places, thus leading to subsequent accidents.
7. Collapse of buildings or structures: Building or structure collapse along with the difficulty to escape from such situations may cause people injuries. The power or gas supply pipelines will also be destroyed, which will cause a further train of accidents.

2.2 *Risk assessment on activities at assembly occupancies*

Based on the hazard identification above and risk indicators in literature (Li, 2009; Tong, 2014), a list of indicators were provided to assess the risk of individual activity. According to most of the basic theories on safety management, these risk indicators fall into four categories, which are indicators relating to people, equipment, environment and safety management. In addition, there are three or four indicators in each category. The total number of items is 15. These 15 items are developed by means of the Delphi expert consensus method. Delphi is a unique method for developing group consensus on what should be included in a subject matter in which precise information is not readily available. Accordingly, I evaluate this method as suitable for my research purpose. In our case, the Delphi method included two rounds of data collection. Several researchers have suggested that a Delphi panel size between 8 and 15 is sufficient (Kim et al., 2013). To guarantee the validity of the Delphi process, 14 experts currently working at universities and doing research in the OHS research field were invited to participate in the development of the indicators. A 5-point Likert scale is used to identify the relevance of each objective. Panel members were asked to rate each item by evaluating its relative importance to assembly occupancy's risk. For this portion of the questionnaire, a 5-point Liker scale (1 – not important, 2 – rarely important, 3 – somewhat important, 4 – moderately important and 5 – very important) was used to quantify the measurement. Finally, 15-item risk indicators were developed for the risk assessment on activities at assembly occupancies.

To get the total risk of each activity at assembly occupancies, Fuzzy Comprehensive Evaluation (FCA) was adopted to calculate the total risk of each activity (Kou,2008). Risk assessment result was divided into five categories, which were very high risk, high risk, medium risk, low risk and ignorable risk.

Table 1. Indicators for risk assessing on activity at assembly occupancies and their priority weight value.

Indicator Category	Weight value	Indicators	Weight value
Indicators relating to equipment	0.2222	Lighting	0.0709
		Traffic condition	0.4557
		Equipment or building reliability	0.2279
		Inflammability of equipment	0.2455
Indicators relating to environment	0.1111	Explosion risk from environment	0.6153
		Risk from weather	0.2922
		Disease transmission risk	0.0925
Indicators relating to people	0.4444	Average people density	0.4513
		Allocation of people	0.2897
		Allocation of people's age	0.1898
		Education level of people	0.0692
Indicators relating to management	0.2222	Alarming systems	0.0601
		Safety instructions	0.1615
		Safety guides	0.2878
		Activity character	0.4905

Each indicator plays different roles in risk assessment; therefore, weight value for each item should be set. This study used Analytic Hierarchy Process to set the weight value of the indicators.

Assessment criteria were provided according to the regulations published and experts' opinions (Table 2).

2.3 Principles of safety management and supervision for various risks

Dynamic safety management and supervision means risk control measurement should be adjusted according to the risk level of various activities. Dynamic safety management and supervision can make sure that the risk is controlled without wasting management resources. Principles of safety management and supervision were developed for each risk level.

3 APPLICATION AND DISCUSSION

3.1 Application of dynamic safety management and supervision to a graduation party

There was a graduation party at a university. Normally, the university assigns more supervisors or guiders to this kind of activities without other special risk control measurements. There were playing and dancing in the graduation party. The number of students taking part in the party was above 2000. The party was held outside in an open area at night.

According to the risk assessment method above, the degree of membership to very high risk, high risk, medium risk, low risk and ignorable risk was 0.3629, 0.0997, 0.1480, 0.0858 and 0.3035, respectively. The total risk of graduation party was medium risk based on the computational method of FCA. So, a special safety management plan was developed on the basis of hazard identification and an emergency plan was set. However, the degree of membership to very high risk was relatively high. So, we should find out the hazard leading to it and take measurements to control the risk to an acceptable level.

3.2 Discussion

Applying dynamic risk control for individual activity at assembly occupancy means adjusting management and supervision according to risk. However, the factors which contribute to the

Table 2. Assessment criteria for indicators.

Indicators	Ignorable risk	Low risk	Medium risk	High risk	Very high risk
Lighting	Above 500 lux.	300–500 lux.	100–300 lux.	10–100 lux.	Below 10 lux.
Traffic condition	Everyone is free to run without affecting each other	Enough space for freely pass	High human flow, but does not affect normal speed	Need to wait in line for passage	Difficult to pass because of narrow road or traffic congestion
Equipment or building reliability	With buildings or structure totally reliable	There is very small probability to collapse	There is small probability to collapse	Similar building or structure has collapsed before	With building or structure easily collapsed (Lin, 2006)
Inflammability of equipment	Noncombustible equipment and building	With equipment and building hard to be combusted	With equipment and building that could be combusted but hard to spread	With equipment and building easily be combusted but hard to spread	With equipment and building easily be combusted and easy for fire spread
Explosion risk from environment	Not any explosive	There is no special activity causing explosion	More explosive accumulated than usual	With activities like setting off fireworks in spacious space	With activities like setting off fireworks with flammable and combustible surroundings
Risk from weather	Sunny	Light rain	Heavy rain and wind	Heavy rain, thunder and lightning	Typhoon, hail and dust storm
Disease transmission risk	No infectious disease	Only a few people caught light infectious disease, such as cold	Epidemic of slight infectious diseases, such as influenza	Serious infectious disease epidemic but not found in local area	There is epidemic of severe infectious diseases in local area
Average people density	Below 0.1 person/m^2	0.3–0.1 person/m^2	0.7–0.3 person/m^2	1–0.7 person/m^2	Above 1 person/m^2
Allocation of people	People scatter without gathering group	People scatter with few small gathering group	People scatter with some gathering group	With several gathering groups of many people	A lot of people gather together in a large group
Allocation of people's age	Totally young people	Mainly young people	Number of people in each age group is even	High proportion of old people or children	Totally old people or children
Education level of people	Most people are well educated	Most people with education above high school	Most people with education above middle school	Most people with education of primary school	Most people with no education
Alarming systems	Enough and used friendly alarming device	Enough alarming device and easy to find	Not enough alarming device	Old or low-reliability alarming device	No alarming device
Safety instructions	With enough and obvious safety signs	With enough but not obvious safety signs	With only a few safety signs in high risk spaces	Without enough or obvious safety signs	No safety sign
Safety guides	Enough guiders with adequate training	Enough guiders with simple and normal training	Not enough guiders, and they take no special training	A few part-time guiders	No guiders
Activity character	With activities which do not lead to mass competition, movement	With activities which may cause the masses to slightly move	With activities which may cause the masses to move under guiding	With activities which may lead to congestion and riots	With activities which may incite the atmosphere of masses and are likely to lead to riots

Table 3. Principles of safety management and supervision for various risks.

Risk	Principles of safety management and supervision with different risk levels.
Very high risk	The activity should not be held until risk is controlled to medium level. Hazard identification, risk assessment and risk control should be done before the activity. Concrete safety management plan and emergency plan should be developed. Emergency exercise should be conducted before activity.
High risk	The activity should not be held unless it is necessary. Hazard identification, risk assessment and risk control should be conducted before the activity. Concrete safety management plan and emergency plan should be developed. Table emergency exercise should be conducted before the activity.
Medium risk	Concrete safety management plan and emergency plan should be developed. Risk should be controlled to as low as reasonable practice.
Low risk	Monitoring the risk should be conducted during the whole process of activity.
Ignorable risk	No special risk control measurement is mandatory.

total risk should also be analyzed to make sure hazards with high risk to be controlled to an acceptable level.

The result of each assessor might be subjective. So, people who conduct risk assessment should be very familiar with the activity and this assessment method. A more specific assessment criterion can also be developed further to solve this problem.

The calculation of total risk is a quite hard work to do manually. A software is suggested to be developed to promote the efficiency of risk assessment.

4 CONCLUSION

This paper has demonstrated the application of the risk assessment as an effective way to adjust safety management and supervision of different activities dynamically. Risk of each individual activity is determined by the 15 indicators falling into the four categories. Total risk was provided by composing the assessment result by the FCA method. Recommendations for future research in this topic are that the assessment criteria should be modified and the assessing indicators should be revised by further applications.

REFERENCES

Kim, K.K., O'Bryan, C.A., Crandall. P.G., Ricke, S.C., &Neal, J.A.Jr., 2013. Identifying baseline food safety training practices for retail delis using the Delphi expert consensus method. *Food Control* 32:55–62.

Kou, L.2009. Study on risk assessment theory and standardization method of assembly occupancies [D]. Beijing: China University of Geosciences (Beijing).

Li, T. 2009. Fuzzy evaluation method and early warning management system design of accident risk assessment in large-scale activities [D]. Beijing: China University of Geosciences (Beijing).

Lin, L.2006.The exploration on safety grade assessment of high occupancy buildings. [D].Chongqing: Chongqing University.

Tong, R.2014. Research on model and method of accident risk assessment in large-scale activities. *China Safety Science Journal* 24 (3):150–155.

Tong. R.2013. *Risk management in large activities accidents-theory & practices* [M]. Beijing: China Labor and Social Security Publishing House.

Hydraulic Engineering IV – Xie (Ed.)

Anaerobic co-digestion of waste activated sludge and waste wine distillate: Effects of polyacrylamide addition and mixing intensity

Lei Zhang, Mingqing Tai & Jianyang Song
School of Civil Engineering, Nanyang Institute of Technology, Nanyang, China

ABSTRACT: We studied the effects of Mixing Intensities (MI) and Polyacrylamide (PAM) addition dosages on the co-digestion of Waste Activated Sludge (WAS) and waste Waste Wine Distillate (WWD) at 35°C and 55°C. The experiments were conducted in eight identical laboratory-scale 51 digesters. Two mixing intensities were used. Results showed that, compare-d to Low Mixing Intensity (LMI), High Mixing Intensity (HMI) did not obviously improve biogas production with the addition of PAM (12 g/kg TS). This was possibly because at the low PAM concentration LMI might be sufficient to provide adequate mixing. With the addition of PAM (24 g/kg TS), a higher Volatile Fatty Acid (VFA) accumulation was observed under LMI conditions. This was possibly because at higher PAM concentrations LMI might not be sufficient to provide adequate mixing. Compared to LMI, the biogas production with HMI at 35°C and 55°C increased by 7.7% and 22.3%, respectively. In addition at 35°C, the difference in effluent Adenosinetriphosphate (ATP) concentrations between of the two mixing intensities is negligible, while at 55°C, while at 55°C, the difference in effluent ATP concentrations between of two mixing intensities is marked.

1 INTRODUCTION

Anaerobic digestion is a biodegradation process, which uses a consortium of natural bacteria to convert a large portion of the organic solids in wastewater into biogas. However, conventional anaerobic digestion of Waste Activated Sludge (WAS) generates only low organic removal and methane production rates. Anaerobic co-digestion of WAS bv adding other materials is an effective method. Its benefits include dilution of potential toxic compounds, synergistic effects of microorganisms and an increased load of biodegradable organic matter (Sosnowski et al., 2003).In the literature many examples of successfully conducted co-digestion processes of different substrates can be found (Kim et al., 2003; Mutro et al., 2004; Gomez et al., 2006). A full scale anaerobic co-digestion with 2000 m³ digestion volume has been used (Zupančičet al., 2008).Hence, it is believed that co-digestion techniques can be widely applied in industry in the near future.

Chemical flocculants are added to enhance sewage sludge sedimentation rate and dewatering before digestion (Chu et al., 2003). Polyacrylamide (PAM) is widely used in wastewater treatment and management. Dentel et al. (2000) demonstrated that the applied PAM may be attached to the conditioned sludge at an amount up to10 g/kg TS. Although the fraction of PAM in the sludge is small, its role has a significant influence on the anaerobic digestion. Although the fraction of PAM in the sludge is small, its role has a significant influence on the anaerobic digestion. El-Mamouni et al., (2002) reported that only a small fraction (0.7%) 0f raw PAM was microbially-mineralize dafter a 38-day incubation under anaerobic conditions. Hence, the characteristics of PAM will persist during anaerobic digestion. However, results reported by Bhunia et al. (2008) suggested that, with PAM addition, a barrier layer is formed surrounding the bacterial cell, which resists diffusion of substrates into the cell, and that the thickness of this barrier layer can be effectively decreased by elevating the mixing intensity.

Mixing plays several roles during anaerobic digestion of sludge, it creates a homogeneous substrate preventing stratification and the formation of a surface crust, and ensures that solids remain in suspension. Moreover, mixing also enables heat transfer, particle size reduction and substrate contact with the microbial community. However, results reported by Vavilin et al (2002) and Vavilin et al. (2005) showed that continuous mixing is not necessary for good performance at higher loading rates and that when organic loading is high, high mixing intensities resulted in acidification and failure of the process, while low mixing intensities were crucial for a successful digestion. When loading is low, mixing intensity bas no significant effect on the process. In summary, appropriate mixing is necessary for presenting substrate to the bacteria, but excessive mixing can reduce biogas production and waste energy. Nevertheless, there is little information about the effect of Mixing Intensity (MI) on anaerobic co-digestion of WAS and WWD in the presence of added PAM. In addition, anaerobic digestion can take place at psychrophilic temperatures below 20°C, but most reactors operate at either mesophilic or thermophilic temperatures, with optima at 35°C and 55°C, respectively (Ward et al. 2008). Furthermore, Gómez et al (2006) studied the impact of mixing on anaerobic co-digestion of sewage sludge and fruit and vegetable fractions of the municipal solid waste. Stroot et al (2001) reported the impact of various mixing conditions on anaerobic co-digestion of municipal solid waste and biosolids. Hence, the impact of mixing conditions on anaerobic co-digestion of flocculated WAS should be studied.

This paper investigates the performance of co-digestion of WAS and WWD in the presence of different concentrations of added PAM and at different MI at both 35°C and 55°C. Biogas and Volatile Fatty Acid (VFA) production, Oxidative and Reductive Potential (ORP), viscosity, and Adenosinetriphosphate (ATP) concentration were measured.

2 METHODS

WAS was taken from the Municipal Wastewater Treatment Plant in Nanyang City, China. WWD was taken from the Alcohol PIant of Nanyang. Seed sludge was taken from thermophilic anaerobically digested sludge. Cationic PAM was used as a flocculant. Its molecular weight, ionic degree and charge density were 12 m III ion. 50–60% and 20%, respectively. The properties of materials are shown in Table 1.

Eight laboratory scale digesters (Dl-D8), were operated at 35 ± 1°C or 55 + 1°C. Each digester had a working volume of 51. Digesters were operated at a Low Mixing Intensity (LMI; 80 rpm) or High Mixing Intensity (HMI; 200 rpm) conditions according to the methods reported by Gomez et al. (2006).

At start-up, 51 seed sludge was transferred to each of the digesters, which were subsequently, kept under anaerobic conditions. From the second day, WAS/WWD mixture (3:1 WAS: WWD v/v; 400 ml) with added PAM was fed to the digesters. Feeding was carried out every two days. Before the feeding the same volume of effluent was removed. All digesters were operated for 50 days. The effluent parameters were measured on at least six samples before the end of the co-digestion performance. The operating conditions of anaerobic co-digestion are shown in Table 2.

A digital pH meter was used for measurement of pH Volatile Fatty Acid (VFA) was measured by gas chromatography, using a HP 6890fFID chromatographer (US). Conductivity, ORP, and viscosity were measured by conductivity meter (Hanna instruments), ORP

Table 1. Properties of materials.

Paramerer	pH	COD(g/l)	TS(g/l)	VS(g/l)	TN(g/l)	TP(g/l)	VFA(g/l)
WAS	6.62	28.6	22.34	14.75	1.42	0.15	169
WWD	4.04	40.7	38.40	30.64	1.75	0.22	547
Seed Sludge	7.72	29.67	58.63	31.37	1.52	0.46	148

Table 2. Anaerobic co-digestion performance conditions.

Anaerobic co-digestion	D1	D2	D3	D4	D5	D6	D7	D8
Operation temperature(°C)	35	35	55	55	35	35	55	55
Mixing intensity	LMI	HMI	LMI	HMI	HMI	LMI	LMI	HMI
PAM addition dosage(g/kg TS)	12	12	12	12	24	24	24	24
SRT(d)	25	25	25	25	25	25	25	25
Performance time	50	50	50	50	50	50	50	50

meter (H198120, Hanna instruments) and viscosity meter (NDJ-1. Shanghai), respectively. Adenosinetriphosphate (ATP) concentration was measured according to the method (Chu et al. 2003).

3   RESULTS AND DISCUSSION

3.1   pH, VFA and alkalinity

The values or pH during co-digestion ranged from 6.81 to 7.53. There was a decrease in effluent pH with increased mixing intensity at 55°C, but at 35°C effluent pH was more or less constant with increased mixing intensity at a PAM concentration of 12 g/kg TS, but showed an increase with mixing intensity at a PAM concentration of 24 g/kg TS.

With the notable exception of co-digestion (D5) with LMI at 35°C in the presence of PAM (24 g/kg TS) the effluent VFA of anaerobic co-digestion at 55°C was higher than that at 35°C with the same PAM addition and mixing intensity. This is in agreement with the results Kaparajuet al., (2008).They found that a higher level of VFA appeared in thermophilic than in mesophilic digestion, since the activity of thermophilic bacteria was higher than that of' mesophilic bacteria. In addition, with the same exception, effluent VFA production with HMI is higher than with LMI at the same PAM addition and temperature. This is maybe due to the fact that that mass transfer is increased with increasing mixing intensity.

Exceptionally. by far the greatest VFA accumulation was observed when PAM addition was 24 g /kg TS at 35°C with LMI. This is possibly because at higher PAM addition, LMI might not be sufficient to provide effective mixing, leading to limitation of mass transfer. Similar results were reported by Veeken et al., (2000), but the VFA accumulation observed here was less than that recorded in their study.

The data of VFA-to-alkalinity ratios indicates that the changes in Ml have no obvious impact on the VFA-to-alkalinity ratios with PAM addition of 12 g/kg TS at 35°C, while there was an obvious change in VFA-to-alkalinity ratios by changing the Ml at 55°C. Carballa et al (2007) showed that the VFA-to-alkalinity ratio of mesophilic anaerobic digestion was almost constant, around 0.15–0.35, while in thermophilic digestion, this ratio was 0.2–0.5. Araya et al. (1999) studied the anaerobic treatment of effluents from an industrial polymers synthesis plant and found that the ratio of VFA to alkalinity was always between 0.5 and 0.75. Van der Stelt et al. (2007) found that ammonia volatilization increased with increasing temperature and mixing, however, at 35°C mixing reduced NH_3 emissions compared to non-mixing, which is related to a reduced crust resistance to gaseous transport at higher temperatures for non-mixing.

The VFA accumulation was probably the result of low microbial level as suggested by Stroot et al. (2001). A correlation between VFA accumulation, ORP and ATP concentration was observed in this study. i.e. a higher VFA accumulation could lead to higher ORP and lower ATP levels. For example, in D5, the VFA was 1704 mg/l, which was the highest among all VFA values; ATP concentration of 6.7 µg/l was the lowest among all ATP values; ORP was the highest of all ORP values.

3.2 Effluent ORP

Chang et al. (1996) suggested that the change in ORP may be attributable to growth of the anaerobic consortium in biosolids. Table 1 and Table 2 show that there was no obvious influence of MI on ORP when the PAM addition was 12 g/kg TS at 35°C, but obvious ORP differences between LMI and HMI are demonstrated when the PAM addition was 12 g/kg TS at 55°C. Moreover, there was an obvious influence of Ml on ORP when PAM addition was 24 g/kg TS at 35°C or 55°C. Among all ORP values, the most negative ORP (−357 mV) appeared with PAM addition of 12 g/kg TS at 55°C and HMI, the least negative ORP (−282 mV) appeared with PAM addition of 24 g/kg TS at 35°C and LMI. This suggests that with addition of 24 g/kg TS at 35°C LMI can inhibit the co-digestion process, whilst with PAM addition of 12 g/kg TS at 55°C HMI does not greatly inhibit the co-digestion process. Hence, it can be concluded that at higher PAM additions a higher mixing intensity is needed for better growth of the anaerobic consortium. This conclusion is similar to that reported by Karim et al. (2005). They considered that mixing becomes more critical during anaerobic digestion of a thicker manure slurry.

3.3 Effluent viscosity

Data on viscosity show that under the same temperature conditions with the same addition of PAM, effluent viscosity at HML drops compared with that at LMI. For example, when PAM addition was 12 g/kg TS, under thermophilic condition, effluent viscosities at HMI and at LMI were 210 and 185 mPa.s, respectively. This illustrates that higher shear can decrease the effluent viscosity during co-digestion. In addition, at the same PAM addition and mixing intensity, the effluent viscosity at 55°C was more than that at 35°C. For example, when PAM addition was 24 g/kg TS with HMI, effluent viscosities at 55°C and 35°C were 220 and 175 mPa.s, respectively. Itävaara et al., (2002) found that the biodegradation of polylactide (PLLA. a polymer, with similar characteristics to PAM) under anaerobic conditions was faster at 52°C than at 37°C, and that temperature is the key parameter affecting the biodegradation rate of PLLA. Therefore we can infer that the viscosity increase with increased temperature is due to the arrangement of degraded end products and residual PAM. This is in agreement with the results of Uyanik et al. 2002. They suggested that when polylactide (PLLA) is exposed to high temperatures, fundamental microstructural changes and molecular rearrangements occur and hydrolysis begins.

Table 3. Characteristics of effluents of co-digestion by addition PAM of 12 g/kg TS.

Digester	pH	ORP(mv)	ATP(μg/l)	Viscosity (mPa.s)	VFA (mg/l)	Alkalinity (mgCaCO₃/l)	VFA/ Alkalinity
D1	7.33	−314	8.2	152	195	593	0.33
D2	7.34	−316	8.4	148	201	662	0.22
D3	7.53	−340	10.5	210	388	1482	0.26
D4	7.17	−357	11.3	185	853	1216	0.70

Table 4. Characteristics of effluents of co-digestion by addition PAM of 24 g/kg TS.

Digester	pH	ORP(mv)	ATP(μg/l)	Viscosity (mPa.s)	VFA (mg/l)	Alkalinity (mgCaCO₃/l)	VFA/ Alkalinity
D5	6.81	−282	6.7	195	1704	2410	0.71
D6	7.02	−296	6.8	175	216	582	0.37
D7	7.36	−317	8.3	235	426	1012	0.42
D8	7.20	−349	9.4	220	654	968	0.68

Data on viscosity show that under the same temperature conditions with the same addition of PAM, effluent viscosity at HMI drops compared with that at LMI. For example, when addition of PAM was 24 g/kg TS at 35°C, the difference in effluent ATP concentrations caused by two mixing intensities was also negligible, while at 55°C, the difference of effluent ATP levels caused by the two mixing intensities was clearly obvious. The increase in ATP concentration means that the biomass activity has increased accordingly. This is in accordance with results of Bhunia et al. (2008) who showed that the activity of microorganisms decreased with increased PAM addition dosage. Similar observations on PAM addition were reported by Uyanik et al. (2002). This phenomenon might be due to the formation of a barrier layer surrounding the bacterial cell that resists diffusion of substrates into the cell. The thickness of this barrier layer can be effectively decreased by adopting adequate mixing intensity during co-digestion, thereby increasing diffusion of substrates into the cell.

Chu et al., (2003) Studied anaerobic digestion of WAS by addition of PAM at 35°C. Their results showed steady effluent ATP concentrations of 3–4 μg/l, and no correlation was observed between the presence of polymer flocculants and microbial activity. Here, in contrast, we observed greater effluent ATP concentrations and a significant correlation between mixing intensity change and ATP concentration, except with PAM addition of 12 and 24 g/kg TS at 35°C. This illustrates that mixing intensity and temperature are the main factors for biomass activity in co-digestion in the presence of added PAM.

3.4 *Effluent ATP*

When addition of PAM was 12 g/kg TS at 35°C, effluent ATP concentrations at LMI and HMI were 8.2 and 8.4 μg/l, respectively. This illustrates that the difference in effluent ATP concentrations caused by the two mixing intensities was negligible, while at 55°C, the difference was much more marked. When addition of PAM was 24 g/kg TS at 35°C, the difference in effluent ATP concentrations caused by two mixing intensities was also negligible. while at 55°C. The difference of effluent ATP levels caused by the two mixing intensities was clearly obvious. The increase in ATP concentration means that the biomass activity has increased accordingly. This is in accordance with results of Bhunia et al. (2008) who showed that the activity of microorganisms decreased with increased PAM addition dosage. Similar observations on PAM addition were reported by Uyanik et al. (2002). This phenomenon might be due to the formation of a barrier layer surrounding the bacterial cell that resists diffusion of substrates into the cell. The thickness of this barrier layer can be effectively decreased by adopting adequate mixing intensity during co-digestion, thereby increasing diffusion of substrates into the cell.

Chu et al., (2003) studied anaerobic digestion of WAS by addition of PAM at 35°C. Their results showed steady effluent ATP concentrations of 3–4 μg/l, and no correlation was observed between the presence of polymer flocculants and microbial activity. Here, in contrast. we observed greater effluent ATP concentrations and a significant cant correlation between mixing intensity change.

3.5 *Biogas production*

Data for biogas production during anaerobic co-digestion under the different conditions is shown in Fig. 1 and Fig. 2. When addition of PAM was 12 g/kg TS at 35°C, biogas production levels at LMI and HMI were 3.75 and 3.791 respectively, while at 55°C, biogas production levels at LMI and HMI were 7.30 and 7.321, respectively. These results illustrate that during anaerobic co-digestion with PAM added at 12 g/kg TS, there is no obvious impact of mixing intensity on biogas production.

When PAM dosage was increased to 24 g/kg TS at 35°C, biogas production levels at LMI and HML were 2.84 and 3.061 respectively. Compared with biogas production at LMI, the biogas production at HMI increased by 7.74%. At 55°C, the biogas production levels at LMI and HMI were 4.16 and 5.101 respectively. Compared with biogas production at LML,

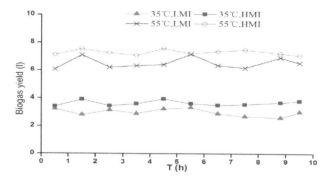

Figure 1.   Biogas yield by PAM addition of 12 g/kg TS.

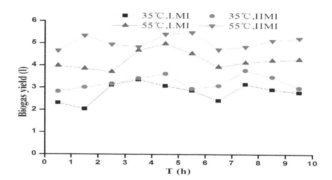

Figure 2.   Biogas yield by addition of 24 gkg TS.

biogas production at HMI increased by 22.3%. Hence, there is an obvious impact of mixing intensity on biogas production when PAM is added at 24 g/kg TS.

4   CONCLUSIONS

Compared with LMI conditions, HMl did not improve the performance of the co-digestion between WAS and WWD in the presence of 12 g PAM/kg TS at 35°C and at 55°C. With the addition of PAM at 12 g/kg TS, under mesophilic conditions the biogas production levels at LMI and HMI were 3.75 and 3.79l respectively, while under thermophilic conditions, the biogas productions levels at LMI and HMI were 7.30 and 7.32 respectively. This was possibly because at low PAM concentrations, LMI is sufficient to provide adequate mixing.

However, the effect of HMI became prominent when co-digestion of WAS and WWD was carried out in the presence of PAM (24 g/kg TS) at both 35°C and 55°C, but especially at 55°C. When PAM was added at 24 g/kg TS at 35°C, compared to LMI, biogas production at HMI increased by 7.74%; while at 55°C, biogas production at HMI increased by 22.3%.

Data for ATP concentrations show that activity of microorganisms decreases with increased PAM addition at the same temperature and mixing intensity.

Based on the findings of this study, it can be concluded that mixing becomes more critical with higher PAM addition. Though the impact of mixing intensity on co-digestion in the presence of added PAM under different temperatures conditions was observed, further research on the impact of mixing intensity on co-digestion in the presence of added PAM is required.

REFERENCES

Bhunia P. Ghangrekar M M 2008. Effects of cationic polymer on performance of UASB reactors treating low strength wastewater[J]. Bioresource Technology, 99, 350–358.

Carbalja M., Omil F., TernesT., et al.,2007. Fate of pharmaceutical and personal care products (PPCPs) during anaerobic digestion of sewage sludge. Water Research 41,2139–2150.

Chang B-V, Zheng JX, Yuan SY.1996. Effects of alternative electron donors and inhibitors on pentachlorophenol dechlorination[J]. Chemosphere 33, 313–320.

Chu CP., Lee D J, Chang B V., You C H., Liao C S., Tay J.H., 2003.Anaerobic digestion of polyelectrolyte flocculated waste activated sludge[J].Chemosphere.53,757–764.

Dentel S K., Chang L L., Raudenbush D R et al., 2000.Influence of synthetic and natural polymer chemistry on biosolids and environment. Water environment Research Foundation, Alexandria, VA.

El-Mamouni, R, Frigon, J.C., Hawari,j., Marroni.,2002. Combining photolysis and bioprocesses for mineralization of high molecular weight polyacrylamides[J]. Biodegradation 13, 221–227.

Gómez X., Cuetos M J., Cara J et al., 2006. Anaerobic co-digestion of primary sludge and fruit and vegetable fraction of the municipal solid wastes. Conductions for mixing and evaluation of the organic loading rate[J]. Renewabie energy 31, 2017–2024.

Itävaara M., Karjomaa S., Selin J F. 2002. Biodegradation of polylactide in aerobic and anaerobic thermophilic conditions[J]. Chemosphere 46, 879–885.

Kaparaju P., Buendia I., Ellegaard L et al., 2008. Effects of mixing on methane production during the thermophilic anaerobic digestion of manure: lab-scale and pilot-scale studies[J]. Bioresource Technology 99, 4919–4928.

Karim K., Hoffmann R., Klasson T., AI-Dahhan M H. Anaerobic digestion of animal waste: Waste strength versus impact of mixing[J]. Bioursource technology 2005, 96:1771–1781.

Kim H W., Han S K., Shin H S., 2003 The optimization of food waste addition as a co-substrate in anaerobic digestion of sewage sludge[J]. Waste management &Research.21,515–526.

Mutro L, Bjornsson B, Matiiasson.2004. Impact of food industrial waste on anaerobic co-digestion of sewage sludge and pig manure [J]. Environ. Manage. 70, 101–107.

Sosnowski P, Wieczorek A, Ledakowicz S.2003. Anaerobic co-digestion of sewage sludge and organic fraction of municipal solid waste[J].Adv.Environ.Res.7,609–616.

Stroot PG, Mcmahon KD, Mackie RI, Rakin 1.2001. Anaerobic co-digestion of municipal solid waste and biosolids under various mixing conditions-I. Digester performance. Water Rcs.35 (7):1804–16.

Uyanik S. Sallis P J, Anderson G K. 2002. The effect of polymer addition on granulation in a anaerobic baffled reactor. Part II: Compartmentalization of bacterial populations [J]. Water Research. 36, 944–955.

Van der Stelt B, Temminghoff E J M, Van Viet P C J et al 2007. Volatilization of ammonia from manure as affected by manure additives, temperature and mixing [J]. Bioresource Technology 98, 3449–3455.

Vavilin V A., Angelidaki I. Anaerobic degradation of solid material: importance of initiation centers for methanogenesis, mixing intensity, and 2D distributed model [J]. Biotechnology and Bioengineering. 89(1), 113–122.

Vavilin V A., Shchelkanov M Y., Rytov S V. 2002. Effect of mass transfer on concentration wave propagation during anaerobic digestion of solid waste[J].Water Research,36,2405–2409.

Veeken AHM, Hamelers BVM. 2000. Effect of substrate-seed mixing intensity, and 2D distributed model [J]. Biotechnol Bioeng. 9(1):l13–122.

Ward A J., Hobbs P J., Holliman P J et al., 2008. Optimisation of the anaerobic digestion of agricultural resources [J]. Bioresource. Technol. 99,7928–7940.

Zupančič G D., Uranjek-Ževert N. Roš M.2008. Full-scale anaerobic co-digestion of organic waste and municipal sludge [J]. Biomass an Bioenergy 32, 162–167.

Hydraulic Engineering IV – Xie (Ed.)
© 2016 Taylor & Francis Group, London, ISBN 978-1-138-02948-4

Application of Risk-Based Inspection (RBI) in atmospheric storage tanks

Yunhua Gong
School of Mechanics, Storage and Transportation Engineering, China University of Petroleum, Beijing, China

ABSTRACT: Storage tank is high-risk facility, which may cause financial loss, environmental pollution and casualties in case of leakage, fir, explosion etc. Under-inspection or over-inspection can occur due to lack of jurisdictional requirements on the inspection interval and method for storage tanks. Reasonable safety inspection is of great significance to ensure the long-term and safe operation for storage tank. Risk-Based Inspection (RBI) is an inspection strategy to make balance between safety and cost. This study applied RBI methodology to optimize the inspection strategy of oil storage tanks. The procedure of RBI according to national standard in China (GB/T 30578–2014 "Risk-Based inspection and evaluation for atmospheric pressure storage tanks") was discussed, and RBI was applied to three crude oil tanks. Risk-based inspection plans for these three tanks were proposed, and suggestions were provided on RBI application in China.

1 INTRODUCTION

Storage tanks are significant and common equipment items in oil, chemical and transportation industry. They are often used to store large amounts of inventory, which are most of the time flammable liquids and sometimes toxic liquids. The hazards from storage tanks can be serious given the large amounts of liquid. On the other hand, storage tanks can cause serious environmental problems when the liquid leak reaches surface waters or underground waters. Another difficulty with floor leaks is that they can go undetected for a long time and can cause serious contamination of the soil or sub-surface water. (Topalis et al., 2012).

Currently, Periodic Internal Inspection is widely used by China's petroleum and chemical industry for the management of crude oil tanks. In China, the code SY/T 5921 (Oil and Gas Storage and Transportation Standardization Technical Committee in China, 2000) has specific requirement for Internal Inspection Interval (INTII): for crude oil tank it is generally 5–7 years; the maximum INTII for the new tank cannot exceed 10 years. However, the Periodic Internal Inspection Method often results in Under-inspection or over-inspection. (Shuai et al., 2012)

RBI helps operators to identify the tanks that do not require frequent internal inspection and repair and avoid the wastage of maintenance and inspection resources, while, at the same time, they can use their resources where it matters: when the risk is high and the inspections are useful. RBI is the developing trend for equipment inspection internationally. The optimal inspection frequency is determined according to its risk exposure, which can be used to avoid any unacceptable risks from under inspection of some items or from over-inspection of the majority of items. (Chang et al., 2005)

This article attempted to incorporate the concepts of RBI with three atmospheric storage tanks and provide suggestions on RBI application in China.

2 METHOD

2.1 *RBI procedure of storage tanks*

The most accepted method on RBI is provided by API 581 "Risk-Based Inspection Base Resource Document" (API, 2008). A standard on RBI of storage tank was developed according to it in China, which is GB/T 30578–2014 "Risk-Based inspection and evaluation for atmospheric pressure storage tanks". (General Administration of quality supervision, inspection and Quarantine of China & China National Standardization Management Committee, 2014) This paper adopted this national standard in case study. The RBI procedure of storage tanks is depicted in Figure1.

2.2 *Risk analysis methodology*

Risk analysis is a critical step of RBI process. Basically there are three approaches to risk assessment: (1) qualitative, (2) quantitative, and (3) semi-quantitative. Qualitative risk analysis is an analysis that uses broad categorizations for probabilities and consequences of failure, while quantitative risk analysis uses logic models depicting combinations of events that could result in severe accidents and physical models depicting the progression of accidents and the transport of a hazardous material to the environment. Semi-quantitative is a term that describes any approach that has aspects derived from both the qualitative and quantitative approaches. (Shishesaz et al., 2013) In this study, GB/T30578–2014 methodology was adopted for risk assessment of storage tanks. In GB/T30578–2014 methodology the failure is defined as loss of containment, and the risk of failure is calculated based on Equation (1):

$$R(t) = F(t) \times C(t) \tag{1}$$

where $F(t)$ stands for Probability of Failure and is a function of time t, and C(t) is consequence of failure. The risk is also a function of time. The probability is calculated by the following Equation:

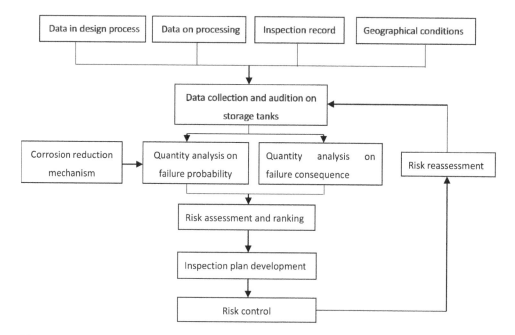

Figure 1.   RBI procedure of atmospheric storage tanks.

$$F(t) = F_G \times D_{f-total} \times F_M \tag{2}$$

where F_G stands for Generic Failure Frequency and is a probability of failure developed for storage tanks based on a large population of component data that does not include the effects of specific damage mechanisms. F_M is a factor which adjusts the generic failure frequencies for differences in process safety management systems. The factor is derived from the results of an evaluation of a facility or operating unit's management systems that affect plant risk. $D_{f-total}$ is an adjustment factor applied to the generic failure frequency to account for damage mechanisms that are active in atmospheric pressure storage tanks.

The consequences are calculated financial-based. The failure analysis of RBI method is the result of the loss of equipment content caused by corrosion degradation, and the result of content loss is usually estimated by the volume and effect of fluid to the environment. The consequence for each storage tank is summarized as follows:

$$FC_{total} = FC_{environ} + FC_{cmd} + FC_{prod} \tag{3}$$

FC_{total} stands for the total financial consequences of potential accident. $FC_{environ}$ is the total financial consequence to environment. FC_{cmd} is the total financial consequences to equipment, and FC_{prod} is the consequences of stop production.

A 5×5 matrix is used for presenting the risk, see Fig. 2. The location of each storage tank on the risk matrix is determined based on the calculated $F(t)$ and FC_{total}. The Probability and Consequence categories can be determined using the guideline given in Table 1.

3 RESULTS AND DISCUSSION

3.1 Results

The tanks are located in a commercial crude oil reserve base with 38 tanks put into use at the end of 2008 and early 2009. Table 2 shows the basic information about these three storage tanks.

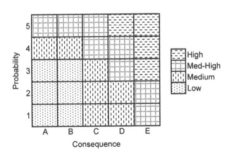

Figure 2. Risk matrix.

Table 1. Numerical values associated with probability and financial-based consequence categories.

Probability Category	F(t)	Consequence Category	FC_{total}
1	$F(t) \le 0.00001$	A	$C(t) \le Q$
2	$0.00001 < F(t) \le 0.0001$	B	$Q < C(t) \le 10Q$
3	$0.0001 < F(t) \le 0.001$	C	$10Q < C(t) \le 100Q$
4	$0.001 < F(t) \le 0.01$	D	$100Q < C(t) \le 1000Q$
5	$0.01 < F(t) \le 1.0$	E	$C(t) > 1000Q$

Note: Q (/Ten thousand Yuan) is the acceptable consequence criteria of certain organization.

Table 2. General information of storage tanks in this study.

Basic parameters of storage tanks in this study

Structure	Storage tank with floating roof	Inventory	Crude oil
Capacity (m³)	100000	Diameter (m)	80
Design storage liquid gravity (kg/m³)	850–920	Maximum operating temperature	Normal temperature
Height (m)	21.8	Design liquid height (m)	20.2
Liquid Operation limit height (m)	2.0–19.5	Liquid safety limit height (m)	2.5–19.0
Thickness in the middle of tank floor (mm)	12	Thickness in the edge of tank floor (mm)	20
Tank wall material	08MnNiVR 16MnR Q235B	Tank floor material	08MnNiVR Q235B

Table 3. Results for probability of failure.

No. of Storage tanks	F_G	$D_{f\text{-total}}$	F_M	Probability of failure	Probability category
G104	0.00072	1.0	0.3516	0.000253125	3
G112	0.00072	6.0	0.3516	0.001518912	4
G124	0.00072	2.0	0.3516	0.000506304	3

Table 4. Risk assessment results.

No. of Storage tanks	Probability category	Consequence category	Risk category
G104	3	C	Medium
G112	4	C	Medium high
G124	3	C	Medium

Table 5. Results for consequence of failure.

No. of Storage tanks	$FC_{environ}$ (RMB)	FC_{cmd} (RMB)	FC_{prod} (RMB)	FC_{total} (RMB)	Consequence category
G104	4,742,089	58,963	1,027,778	5,828,830	C
G112	4,742,089	58,963	1,027,778	5,828,830	C
G124	4,742,089	58,963	1,027,778	5,828,830	C

Note: Due to the same size and location, these three storage tanks share the same financial consequences.

According to the guideline in GB/T30578–2014, the probability of failure, total financial consequence of potential accident and the final risk assessment are displayed in Table 3, 4 and 5.

Based on these three case studies, we can obtain valuable results as follows:

1. Normally, storage tanks holds relatively high risk because of the severe consequences in case of failure.
2. Storage tanks' risk is highly influenced by safety management because of the weigh value of management in probability calculation.

Table 6. Inspection plan of each storage tank.

No. of Storage tanks	Inspection method	Next inspection time
G112	Open tank inspection; Online inspection.	now
G124	Open tank inspection; Online inspection.	Conduct open tank inspection within 1 year. Conduct online inspection during next year.
G104	Open tank inspection	Within 8 years.

3. Some of storage tank risks increase quickly. Risk of No. G124 storage tank will reach medium high in one year if no action is taken. However, risk of No G104 has a relatively slow increasing trend.

Inspection plans drawn up according to risk assessment are shown in Table 6.

3.2 *Discussion*

In the process of applying RBI to storage tank, it is found that due to the same failure consequence of storage tank in the same site under the same safety management system, these storage tanks have a high probability to share the same risk category, which leads to the difficulty to distinguish inspection plans among them. Therefore, the developing trend of risk is of great significance in application of RBI. That is to say, some storage tanks have risks increasing more rapidly than others. Hence, inspection plan should consider not only the risk category but also take the developing trend of risk into account.

Effective and reliable data is critical to RBI analysis. Due to lack of data, some parameters take the conservative estimate in this study. In this case, results may exist certain deviation. Therefore, enterprises should strengthen the work of data collection to take advantage of RBI methodology effectively.

4 CONCLUSION

This paper reviews RBI methodology designed for aboveground atmospheric pressure storage tanks. The RBI methodology includes a risk assessment process for storage tanks to rank inspections for a variety of tanks. It is of great value to optimize the inspection plan of oil storage tank. It is essential that high-level risk storage tank be properly managed, and that sufficient resources (manpower and budget) be allocated to this effort.

An integral inspection methodology on storage tanks should consist not only of an efficient inspection strategy and reasonable inspection planning, but also of reliable inspection methods. Hence, further research could be conducted on improving the effectiveness of inspection methods to enhance RBI applicability. And, to obtain precise risk assessment result, building up a more detailed database of storage tanks is indispensable for conducting RBI in China.

ACKNOWLEDGEMENT

This paper was supported by Science Foundation of China University of Petroleum, Beijing (No. ZX20150085).

REFERENCES

API 581 American Petroleum Institute. 2008. *Risk-based inspection technology*. Washington, D.C.: API Publishing Services.

Chang, M.K., Chang, R.R., Shua, C.M., & Lin, K.N. (2005). Application of risk based inspection in refinery and processing piping. *Journal of Loss Prevention in the Process Industries* 18:397–402.

GB/T 30578–2014 General Administration of quality supervision, inspection and Quarantine of China, China National Standardization Management Committee. 2014. *Risk-Based inspection and evaluation for atmospheric pressure storage tanks.* Beijing: China Standard Press.

SY/T 5921 Oil and Gas Storage and Transportation Standardization Technical Committee in China. 2000. *Code for repair of vertical cylindrical weld steel crude oil tanks.* Beijing: Petroleum Industry Press.

Shishesaz M.R., Bajestani M.N., Hashemi S.J. and Shekari E. 2013. Comparison of API 510 pressure vessels inspection planning with API 581 risk-based inspection planning approaches. *International Journal of Pressure Vessels and Piping* 111–112:202–208.

Shuai J., Han K.J., Xu X.R. 2012. Risk-based inspection for large-scale crude oil tanks. *Journal of Loss Prevention in the Process Industries* 25: 166–175.

Topalis P., Korneliussen G., Hermanrud J. and STeo Y. 2012. Risk Based Inspection Methodology and Software Applied to Atmospheric Storage Tanks. *Journal of Physics: Conference Series 364*:1–13.

Hydraulic Engineering IV – Xie (Ed.)
© 2016 Taylor & Francis Group, London, ISBN 978-1-138-02948-4

Calculation of CO_2 emission factors from thermal power industry based on environmental statistics 2014

Ying Zhou & Zhansheng Zhang
The Center for Climate and Environmental Policy, Chinese Academy for Environmental Planning, Beijing, P.R. China

Junxia Wang & Jiong Zhou
State Environmental Protection Key Laboratory of Quality Control in Environmental Monitoring, China National Environmental Monitoring Centre, Beijing, P.R. China

ABSTRACT: In order to simultaneously control air pollution and greenhouse gas emission, China has carried out a lot of researches on co-control of emission of major air pollutants and carbon dioxide (CO_2), which is gradually developing toward being quantitative, microscopic, and technical. Therefore, the emission factors of greenhouse gas and the generation and emission factors of air pollutants become the required parameters in model analysis and policy research. At present, the relatively well-established generation and emission factors of CO_2 and air pollutants in the country and abroad stand on their own feet and are hard to be integrated into the same manifestation mode, mainly for lack of systematic CO_2 generation and emission factor with products (or raw materials) as the unit. Based on relevant data in environmental statistics 2014 concerning thermal power generating units, and with pollutant generation and emission factors as the reference, this paper calculates the CO_2 emission factor of each type of thermal power generating units in China. Furthermore, it puts forward policy suggestions to further improve the environmental statistics system and synergic control of thermal power enterprises' CO_2 and air pollutant emissions.

1 INTRODUCTION

Since the reform and opening-up, China has attained great achievements in economic construction, while various environmental problems are looming large, and particularly, air pollution and greenhouse gas emission have become a bottleneck in deep social and economic development. As air pollution and greenhouse gas emission have the same source, that is, both of them are caused by emission from fossil fuel burning; therefore, controlling air pollution and reducing greenhouse gas emission should be consistent in action (Ding, 2009). Efficient co-control of emission of greenhouse gases and normal pollutants has become a hot research topic at present (Wang, 2010). The research of synergic control is now gradually developing toward being quantitative, microscopic, and technical, thus gradually focusing on the synergic effect of regions, industries, and even specific enterprises in the process of technical progress; therefore, the generation and emission factors of air pollutants and greenhouse gases have become the required parameters in model analysis and policy research. At present, the relatively well-established generation and emission factors of carbon dioxide (CO_2) and air pollutants in the country and abroad stand on their own feet and are hard to be integrated into the same manifestation mode, due to lack of systematic CO_2 generation and emission factor with a certain product (or material) as the unit. Despite that, there are a few data available in the process of model design, most of the time it's applying the default emission factors abroad, which is not in line with China's current status of technological development, and besides, the data are not complete and systemic, which becomes a major shortcoming in model research and direct case analysis of China's synergic control.

Greenhouse gas emission factors mainly refer to technical manuals recommended by various international organizations. Plenty of research has been made around the world on greenhouse gas emission factors, and systematic theories and methods have been developed (IPCC, 2007) (WRI/WBCSD, 2005). In order to draw up a national greenhouse gas list, in the process of drawing up the first (in 1994) and the second (in 2005) edition of National GHG Emission Inventory, China confirmed emission factors that were more in line with the actual situation of China by means of a key enterprise survey, material balance, actual measurement, and so on, based on IPCC calculation method systems (National Climate Change Coping and Coordinating Group Office, 2007; Department of Climate Change of National Development and Reform Commission, 2014; Department of Climate Change of National Development and Reform Commission, 2014). Such research mainly proceeds from a macroperspective, considers different types of fuels consumed aggregately in the industries, and then calculates the CO_2 emissions from using different fuels, which is easy for each enterprise to calculate CO_2 emissions generated from energy burning, but cannot break down into the specific process, raw material, product, and scale.

In power generation industry, at the regional-scale power grid, electric power emission factor is most widely applied. In order to promote the development of more Clean Development Mechanism (CDM) projects in line with international rules and China's key areas, experts from Department of Climate Change of the National Development and Reform Commission had conducted research to determine year-by-year line-based emission factors for China's regional power grid. Meanwhile, even specific to different regions, these CO_2 emission factors have considered all power generation modes in each region, which are very simple and accurate for electric power users but is of little significance to specific power generation enterprises. From the perspective of scientific research, there have also been some researches featuring the actual measurement in samples of small, typical enterprises' greenhouse gas emissions, for instance, CO_2 and NO_x emission factors per unit power generation were obtained after monitoring 30 representative thermal power units in power industry (Wu, 2010). These CO_2 emission factors have important significance to this research, but due to restrictions in budget and project area, as well as limited number and type of enterprises selected, such CO_2 emission factors have certain limitations.

This paper will utilize facility-level database in environmental statistics 2014 to determine the generation and emission factors of CO_2 and air pollutants in a typical industry in China in the context of producing the same product, using the same raw materials with same size, thereby providing basic data and technical information support for synergic control of CO_2 and air pollutants (mainly including SO_2, NO_x, soot, and dust).

2 DATA

The annual environmental statistic work not only includes pollution sources, pollutant generation amount and emissions, but also production-related information such as energy consumption, raw & auxiliary materials consumption, production, and also involves all emission sources in thermal power industry. In view of the fact that greenhouse gas and air pollutant emissions have the same sources and are simultaneous, while providing main sources' pollutant emissions, the environmental statistic work at present can also meet the basic data requirements of all thermal power enterprises and even thermal power units for CO_2 emission calculation.

3 METHOD

3.1 CO_2 calculation method

CO_2 emission calculation method and principles mainly adhere to IPCC guidelines for National Greenhouse Gas Emission Inventories (2006) and IPCC Good Practice Guidance and Uncertainty Management in National Greenhouse Gas Inventories and meanwhile refer to compilation method of the People's Republic of China Initial National Communication on

Climate Change (National Climate Change Coping and Coordinating Group Office, 2007), compilation method of The People's Republic of China Second National Communication on Climate Change (National Climate Change Coping and Coordinating Group Office, 2014a) and Research of China's Greenhouse Gas Inventories 2008 (Department of Climate Change of National Development and Reform Commission, 2014b), and then thermal power enterprises' CO_2 emissions calculation formula 1 is determined:

$$E_i = \sum_j \delta_j \times Q_j \times O_j \times D_j \qquad (1)$$

Therein, E_i is the CO_2 emissions of enterprise i (t), δ_j is the conversion factor, converts thermal unit into ton; Q_j is the carbon emission factor of fossil energy j (kgC/106 kJ); O_j is the carbon oxidation rate of fossil energy j; D_j is the consumed amount of energy j; j is the different energy types, for example, raw coal, cleaned coal, washed coal, and so on.

3.2 Unit of emission factor

In China's First National Pollution Source Census in the year of 2007, according to the sequence of national economy industry code (sub-class), the pollutant generation and emission coefficients per unit of product (raw material) were given under the "four 'same' combination" (classified as per the same kind of product, the same raw material, the same process, and the same scale of production), based on which the first national pollution source census, Manual of Industrial Pollution Sources' Pollutant Generation and Emission Coefficients, was compiled (Data Compilation Committee of the First National Pollution Source Census, 2011), at present, the pollutant generation and emission coefficients of each industry in China mainly follow this manual. Therein, Manual of Pollutant Generation & Emission Coefficients in thermal power industry provided the pollutant generation coefficient and pollutant emission coefficient in the production process of electric energy products, electric energy and thermal energy products, and so on in the thermal power generation industry (industry code 4411), and the unit is ton (kg, g)/ton-material, of which the material refers to fuel.

By and large, thermal power enterprises in China have not implemented effective CO_2 removal technology at present, and therefore, CO_2 generation factor is the same with CO_2 emission factor. Analysis on data reported by thermal power enterprises in environmental statistics reveals that the data related to fuel property (e.g., average lower heating value [average LHV], carbon content, and so on) are incomplete, also with high error; if g/ton-fuel is adopted as the unit of emission factors, the error would be very big. Therefore, the unit of CO_2 emission factor is hereby determined as g/kwh, which can improve the accuracy of the results obtained. So, calculation of formula 2 for thermal power enterprises' CO_2 emission factor is as follows:

Emission factor thermal power = CO_2 *emissions/amount of power generated*   (2)

3.3 Pollutant emission factors

As Manual of Pollutant Generation & Emission Coefficients in Thermal Power Industry has listed a variety of pollutant generation and emission factors for each type of units, as well as plenty of desulfurization and denitrification methods and relatively complex data types, they will not be listed herein one by one and shall only have the corresponding CO_2 emission factors listed.

4 DATA PROCESSING AND ANALYSIS

4.1 Type division of thermal power units

Thermal power units 2014 are classified according to the "four 'same' combination" dividing principle in the Manual of Pollutant Generation and Emission Coefficients in Thermal

Power Industry. However, in environmental statistics 2014, there is no statistics concerning the type of thermal power units, so, in this research, division of units is done by only considering the combination and classification as per the same kind of product, the same raw material, and the same scale of production. According to actual statistics 2014 concerning the scale of thermal power generating units, thermal power units are here divided into 11 types, as shown in Table 1.

Through analysis on relevant data of thermal power generating units in environmental statistics, 4145 effective unit data have been obtained, and the effective number of each type of units is shown in Table 1. Therein, the number of A1-type coal-fired units and oil-fired units is relatively small.

4.2 Calculation of CO_2 emission coefficients

After calculating the CO_2 emission factors of thermal power units with effective information, the paper carried out a normality test and outlier analysis for relevant data on each type of unit and then confirms the final emission factors:

4.2.1 Normality test
The K–S test was performed to check the main parameters and to calculate emission factors of each type of unit in thermal power enterprises, which shows that most parameters and all types of emission factors disaccord with the normality test, and only part of the carbon content data was reported in accordance with the normal distribution. Therefore, in the subsequent outlier analysis and final result determination, data processing will be carried out as per normal distribution and non-normal distribution data, respectively.

4.2.2 Outlier analysis
A simple analysis on the data reveals that, as the data adopted this time could not be checked one by one, the data quality is relatively poor, and meanwhile, these data are hard to be corrected with empirical values. Thus, a box plot is to be used to infer the outliers, and the outliers besides [Q1− 3 × IQR, Q3+3 × IQR] are to be removed.

4.3 Calculation results

Through above analysis, we can obtain the average LHV and carbon content of each type of unit using coal in thermal power industry and obtain the CO_2 emission factor of each type of unit, as shown in Table 2.

Table 1. Number of each type of units available (set).

Type of units	Scale of thermal power units	Number of units
A1	≧50 MW pulverized coal furnace	75
A2	450–749 MW pulverized coal furnace	477
A3	250–449 MW pulverized coal furnace or circulating fluidized bed boiler	877
A4	150–249 MW pulverized coal furnace or circulating fluidized bed boiler	276
A5	75–149 MW pulverized coal furnace or circulating fluidized bed boiler	276
A6	35–74 MW pulverized coal furnace or circulating fluidized bed boiler	150
A7	20–34 MW pulverized coal furnace or circulating fluidized bed boiler	365
A8	9–19 MW pulverized coal furnace, grate-fired furnace, or circulating fluidized bed boiler	706
A9	All ≤8 MW coal-fired boilers	480
B	Oil-fired boiler	45
C	Gas-fired boiler	418

Table 2. CO_2 emission factor of each type of units in thermal power industry (g/kwh).

Type of units	Average LHV	Carbon content	CO_2 emission factor
A1	20796 (3095.29)	47.70 (±8.33)	758.10 (55.20)
A2	19723 (3657)	46.02 (10.89)	836.57 (137.40)
A3	18983.43 (4203.79)	45 (10.89)	892.29 (170.02)
A4	17380.38 (5119.92)	45.70 (11.34)	993.58 (291.14)
A5	18756 (6881.23)	44.26 (±11.45)	1009.43 (370.13)
A6	18753 (8863.25)	46.18 (±12.11)	1077.71 (479.09)
A7	19241.50 (7524.35)	47.21 (±12.11)	1012.06 (443.09)
A8	16210 (8336.5)	41 (14)	1086.42 (699.9)
A9	15325 (8747.57)	42.67 (14.4)	1074.90 (1024.78)
B	–	–	427.60 (794.39)
C	–	–	668.45 (650.67)

5 CONCLUSIONS AND SUGGESTIONS

5.1 Conclusions

1. Many thermal power enterprises also use a small amount of other energies, while using one main source of energy, whereas the generation of SO_2 and NO_x depends mainly on coal use, and few amounts of SO_2 and NO_x are generated from fuel oil and various kinds of gases. Nevertheless, in the process of analyzing CO_2 emission factors, all the energies and fossil energies burned in the units shall be taken into consideration, analyzing their consumption, average LHV, carbon content, and so on. Therefore, it is hard for the CO_2 emission factor to become a functional form relevant to a certain index, like the SO_2 generation and emission factor.
2. CO_2 emission factor of oil-fired and gas-fired units is significantly lower than that of each type of coal-fired unit, whereas difference of which is smaller than the difference in SO_2 generation and emission factor. Therefore, when using oil-fired and gas-fired units, given the same amount of power generation, its CO_2 emissions will be slightly lower than each type of coal-fired units, but can significantly reduce SO_2 generation & emission.
3. On the whole, average LHV and carbon content of domestic thermal power units using coal are lower than the national average, which accords with the suggestion that Chinese thermal power enterprises at a comparatively high pollution treatment level use relatively poor-quality coal, whereas the coal used by thermal power enterprises generally takes on a trend that the higher the installed capacity, the higher the coal-using average LHV and carbon content, which is related to (a) highly installed capacity units' requirement for coal quality and (b) large power generation enterprises' stable and high-quality coal supply channels.

5.2 Suggestions

1. It is suggested that, in the process of carrying out corresponding researches on synergic control, particularly in bottom-up model design (e.g., LEAP), thermal power units' CO_2 emission factors calculated herein are more in line with China's actual situation than the previous default emission factors from abroad, thereby making research results more in line with the general situation of China's thermal power industry.
2. It is suggested that, on the basis of the generation and emission factors of CO_2 and air pollutants, analysis and confirmation of the positive or negative synergic effect of key emission reduction measures in key industries among multi-factors and, combining with the existing typical industry policies, synergic control potential technologies, and the technical measures suggested to be eliminated, be studied.

3. As the power generation plants' unit type and fuel nature are not key review data in environmental statistics, therefore, the integrity and accuracy of which is relatively low, and it is suggested that, in the future data reporting, the reporting guidance be reinforced and rechecked with regard to such indexes to improve the index accuracy.

REFERENCES

Data Compilation Committee of the First National Pollution Source Census, Manual of Pollutant Generation & Emission Coefficients in Pollution Source Census, China Environmental Science Press, 2011.

Department of Climate Change of National Development and Reform Commission, Research of China's Greenhouse Gas Inventories (2005), China Environmental Science Press, 2014a.

Department of Climate Change of National Development and Reform Commission, Research of China's Greenhouse Gas Inventories 2008, China Planning Press, 2014b.

Ding Yihui, Li Qiaoping, Liu Yanju, et al., Air Pollution and Climate Change, Meteorological Monthly, vol.35, no.3, pp.3–14, 2009.

IPCC, 2006 IPCC Guidelines for National Greenhouse Gas Inventories, http://www.ipcc-nggip.iges. or.jp/public/2006 gl/index.html, 2007.

National Climate Change Coping and Coordinating Group Office, Research of China's Greenhouse Gas Inventories, China Environmental Science Press, 2007.

Wang Jinnan, Ning Miao, Yan Gang, et al., Implementing Climate - friendly Strategy for Air Pollution Prevention and Control, China Soft Science, no.10, 10, pp.28–36, 2010.

WRI/WBCSD, CO_2 Emission Reduction Protocol in Cement Industry, Statistical and Reporting Standard for CO_2 Emission in Cement Industry, 2005.

Wu Xiaowei, Zhu Fahua, Yang Jintian, et al., Measurements of Emission Factors of Greenhouse Gas from Thermal Power Plants in China, Research of Environmental Sciences, vol.23, no.2, pp.170–176, 2010.

Hydraulic Engineering IV – Xie (Ed.)
© 2016 Taylor & Francis Group, London, ISBN 978-1-138-02948-4

Effects of different fire floors on the evacuation of high-rise buildings

Rongshui Qin, Xiaoxiao Dong, Chao Song, Xiaojing Hou & Jiping Zhu
State Key Laboratory of Fire Science, University of Science and Technology of China, Hefei, Anhui, China

ABSTRACT: The large number of persons, long evacuation distance, as well as the impact of the special circumstances of fire often lead to a phenomenon such as blind conformity, the bottleneck at the crowded stampede, and even wrong evacuation routes during the high-rise buildings' evacuation. So a high-level scientific evacuation plan is extremely necessary. The existing evacuation model and common business evacuation software such as pathfinder and EVACENT mostly belong to the 'pre-model', setting a critical floor before simulations, that is, people below it use stairs to evacuate and the others use elevators, and repeatedly change the floor in order to ultimately obtain an approximate optimal floor that ensures that the overall evacuation time is the shortest, which greatly reduces the computational efficiency. Considering the limitations of commercial software in evacuation simulation, in our previous work, the stairs and elevators' collaborative evacuation model was developed to quickly generate the optimal evacuation routes through only one computation by using the two factors in collaboration. This article expands the stairs and elevators collaborative evacuation model to research the influence of different fire floors on the evacuation routes in high-rise buildings and then dynamically adjust the evacuation routes to ensure that the total evacuation time is minimum.

1 INTRODUCTION

With China's process of urbanization continuing to accelerate in recent years, investment on construction of high-rise buildings is increasing rapidly, and the landmark high-rise buildings are springing up one after another (Ding et al. 2015). Due to the fact that the high-rise buildings are being built higher and higher, they can accommodate increasing number of people. The required overall evacuation time is thus increasing in fire accidents. Therefore, greater attention has been focused on how to reduce the emergency evacuation time in high-rise buildings using the current evacuation facilities worldwide (Heyes et al. 2012). In order to study the characteristics of evacuation of occupants, scholars put forward a variety of evacuation models, such as network evacuation model (Chalmet et al. 1982; Kisko et al. 1985), lattice gas model (Shang et al. 2015), cellular automata model (Ma et al. 2012), fine discrete model (Cao et al. 2015), and so on. These models have advantages and disadvantages, but for high-rise buildings, in the case of real-time evacuation strategy research, the network evacuation model has a unique advantage of not having to consider the impact of their behavior on the groups and individuals; instead, it can take the same room or person in the region as a whole into consideration. Hence, the network evacuation model has a natural advantage in reducing the computation time.

Our previous work (Dong et al. 2015; Zhu et al. 2013; Dong et al. 2014) has developed a high-rise building collaborative evacuation model using stairs and elevators to produce real-time evacuation path due to a dynamic evacuation network. It has studied the influence of the number of persons to evacuate, and the elevator load on mixed evacuation process, and also considered several factors such as extra evacuees in source nodes on evacuation process. However, they did not consider the effects of the impassable path which is caused by the smoke of the fire on the evacuation process. And in this paper, we use the Fire Dynamics Stimulator (FDS) to calculate the stairs' impassable moment caused by the smoke that spreads from the fire room;then we will study the effects of the different fire floors on the overall evacuation time based on our model.

2   DESCRIPTION OF COLLABORATIVE EVACUATION MODEL

The real-time dynamic collaborative evacuation model can quickly adjust to the evacuation routes based on the current situation if any route is blocked by smoke or other reasons at any time in the evacuation process, which means the time of travel to the corresponding arc in the network graph of the building becomes infinite. The failure arc and the failure moment can be specified and the adjusted evacuation routes ensure that people evacuate to the outside safely in the shortest time.

We use the network graph of a building with five floors in Figure 1 to describe the algorithm of the real-time dynamic network model. Under normal circumstances, the program is based on the minimum arrival time in order to search the shortest time path, so that the evacuees can evacuate on the basis of the best proportion of stairs and elevators evacuation. Assuming that at the fifth time step (25 s) in the evacuation process, the stairs between the second and third floors are impassable due to the spread of smoke, and the corresponding arc between nodes 12 and 16 in the network graph fails suddenly in the program; it means that the travel time of the arc suddenly becomes infinite. In order to record the distribution of evacuees in the network graph and the moment in which the failure of the arc occurs, the function of the addnode () is used to insert some new nodes to record the evacuees' positions in the network graph. At the arc failure time, the program first generates a new evacuation

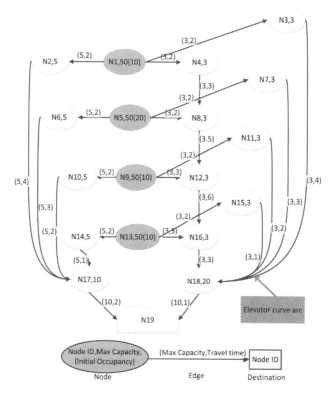

Figure 1.   Network diagram of collaborative evacuation model.

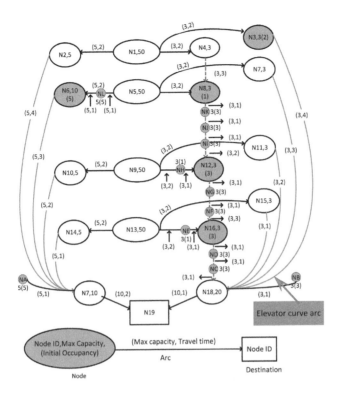

Figure 2. The new network diagram of evacuation at the fifth time step (25 s).

network, whereas Figure 2 represents the new evacuation network diagram recorded by the program at the moment of the failure of arcs 12–16. The initial property values of all the nodes and arcs have been marked on the graph. Comparing Figures 1 and 2, NA to NL and N3, N6, N8, N12, and N16 are the new source nodes, the representatives of the distribution of people at the fifth time step (25 s) in these nodes. Based on the new evacuation situation, the algorithm search for the optimal evacuation routes using stairs and elevators collaborates until all the evacuees evacuate to the safe areas.

3 CASE STUDY

This paper takes the Hiroshima residential building fire case in Japan as example (Sekizawa et al. 1999). The dynamic collaborative evacuation model of this apartment contains 101 nodes and 157 arcs. With 120 people set on each floor, a total of 2400 people are to be evacuated in the simulation project. Using FDS, a CFD (Computational Fluid Dynamics) software, to simulate the situation of fire on different rooms, we can get the timetable for the spread of smoke to the stairwell. Nine fire scenes were developed, and their conditions are listed below (Table 1). There are three types of rooms designed: room of the first type is far away from one staircase but near to another; the second type is between two of the staircases; and the last one is far away from either of the staircases. The positions of the three types of rooms are shown in Figure 3. Three sources of fire power are designed for 1.5MW, 3MW, and 6MW, and the smoke exhaust rate is set as 60 m³/(hm²).

We can acquire the dangerous time of the evacuation path of staircase by using FDS to simulate each case in Table 1. The dangerous time acquired by FDS is the failure time of the staircase of the corresponding floor. We use the high-rise building real-time dynamic evacuation model to disconnect this arc of the path at current time and re-plan the evacuation

Table 1. Fire conditions and dangerous timetables.

Fire room	Case	Fire power (MW)	Dangerous time (s)
Room 1	Case 1	1.5	150
(Close to one staircase)	Case 2	3	125
	Case 3	6	95
Room 2	Case 4	1.5	245
(Between the two staircases)	Case 5	3	140
	Case 6	6	125
Room 3	Case 7	1.5	275
(Far from the two staircases)	Case 8	3	200
	Case 9	6	170

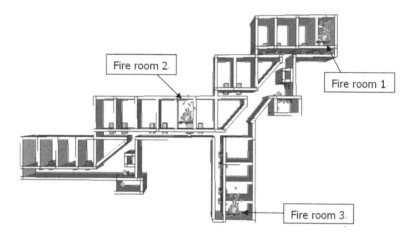

Figure 3. The schematic of a fire room.

route. And then we can get the path planning timetable for the overall evacuation time and the number of people to be evacuated through the elevators or stairs.

Figure 4(a)–(c) shows the influence of different fire floors on the overall evacuation time, whereas (d)–(f) shows the influence of different fire floors on the number of people to be evacuate by the elevator.

The result reveals that a critical floor level exists in every case, and its number alters after the conditions change. When the fire happens under the critical floor (including the critical floor), the total evacuation time and the number of people to be evacuated by the elevator increase with the decrease of the floor level, and it requires more time to evacuate in these scenes. That is to say, the failure of the staircase leads to the adjusting of the evacuation path dynamically. On the other hand, the overall evacuation time and the number of people to be evacuated by elevators almost remain constant when the fire happens above the critical floor.

Figure 5 shows that the lower the fire floor is, the longer the overall evacuation time will be. This indicates certain evacuation scenarios which are more dangerous. Furthermore, when the fire floor is fixed, the greater the fire power is, the longer is the overall evacuation time. This is because the fire power becomes greater, and the failure time of the stairwell becomes earlier. At the same time, there are an increasing numbers of people who have not evacuated, and the stairwell gets crowded obviously, which leads to the adjustment of the optimal evacuation paths and prolonging of overall evacuation time. Figure 5 (d) shows that there is no threat to evacuation of people when the fire happens above the critical floor, so there is no need to adjust the evacuation route dynamically.

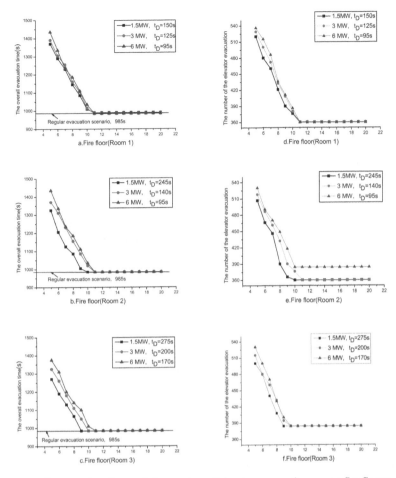

Figure 4. Overall evacuation time and the number of elevator evacuation versus fire floors.

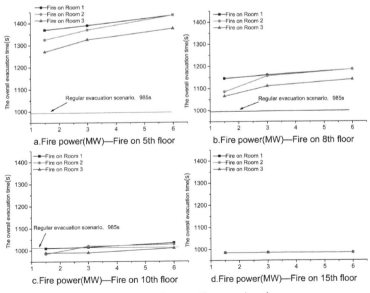

Figure 5. The influence of fire power on the overall evacuation time.

4 CONCLUSION

The real-time dynamic collaborative evacuation model can quickly adjust to the evacuation routes based on different fire floors at any time in the evacuation process to ensure that the overall evacuation time is minimal. According to the model, a critical floor level is derived. When the fire happens above that level, the overall evacuation time and the number of people to be evacuated from elevators do not change. When the fire happens under the critical floor, the time and number increase approximately in a linear manner with the decrease of the floor.

ACKNOWLEDGMENTS

This work was sponsored by the National Basic Research Program of China (No. 2012CB719705), Jiangsu Science-Technology Support Plan Projects (No. BE2012671), and the Fundamental Research Funds for the Central Universities (No. WK2320000033).

REFERENCES

Cao S, Song W, Lv W, et al. A multi-grid model for pedestrian evacuation in a room without visibility[J]. *Physica A: Statistical Mechanics and its Applications*, 2015, 436: 45–61.

Chalmet, L.G., R.L. Francis, and P.B. Saunders, Network Models for Building Evacuation. *Management Science,* 1982. 28(1): p. 86–105.

Ding Y, Yang L, Weng F, et al. Investigation of combined stairs elevators evacuation strategies for high rise buildings based on simulation[J]. *Simulation Modeling Practice and Theory*, 2015, 53: 60–73.

Heyes, E., M. Spearpoint. Lifts for evacuation—human behavior considerations[J]. *Fire andMaterials,* 2012, 36(4): 297–308.

Jiping Zhu, Yunlong Wang, Xiaoxiao Dong. Real-time evacuation path: unexpectedly increasednumbers of evacuees in source node. *Advanced Material Research*, Vols.779–780 (2013) pp 1060–1065.

Kisko, T.M. and R.L. Francis, Evacnet+ - a Computer-Program to Determine Optimal BuildingEvacuation Plans. *Fire Safety Journal*, 1985. 9(2): p. 211–220.

Ma J, Lo S, Song W. Cellular automaton modeling approach for optimum ultra high-rise building evacuation design, *Fire Safety[J]* 54 (2012) 57–66.

Sekizawa A, Ebihara M, Notake H, et al. Occupants' behavior in response to the high-rise apartments fire in Hiroshima City[J]. *Fire and Materials*, 1999, 23(6): 297–303.

Shang H Y, Huang H J, Zhang Y M. An extended mobile lattice gas model allowing pedestrian step size variable[J]. *Physica A: Statistical Mechanics and its Applications*, 2015, 424: 283–293.

Xiaoxiao Dong, Chao Song, Rongshui Qin, Jiping Zhu. Study on High-rise Building Collaborative Evacuation Using Stairs and Elevators Based on Multi Constrained Conditions. *ASPACC 2015–10th Asia-Pacific Conference on Combustion*, 2015.

Xiaoxiao Dong, Yunlong Wang, Jiping Zhu, Chao Song. Influence of evacuees numberandelevator load on mixed evacuation process. *Hydraulic Engineering II - Proceedings of the 2nd SREEConference on Hydraulic Engineering, CHE 2013*, P 211–216, 2014.

Hydraulic Engineering IV – Xie (Ed.)
© 2016 Taylor & Francis Group, London, ISBN 978-1-138-02948-4

Nutrient and heavy-metal contamination of dredged sediment from Taihu Lake

Y. Lin, H.X. Xiong, W. Huang & C.B. Liu
Key Laboratory of Environmental Protection in Water Transport Engineering, Tianjin Research Institute for Water Transport Engineering, Ministry of Transport, Tianjin, China

ABSTRACT: A total of 15 dredged sediments' samples from Taihu Lake were obtained and tested. The nutrient and heavy-metal pollution characteristics were analyzed. Geoaccumulation index (I_{geo}) was applied for the assessment of heavy-metal pollution in dredged sediments. The research results showed that the pH, organic matter, and available P of dredged sediment from Taihu Lake were 7.28, 25.8%, and 8.41 ppm and 89.97 ppm, respectively, which were moderately suitable for plant's growth; the available N and K contents were 16.96 ppm and −20.11 ppm, which were too low for planting, so extra sources of nitrogen and potassium should be added into the sediment in order to meet the requirements for plant's growth while recycling. Based on the I_{geo} of target of heavy metals, the sediment collected from the study area can be approximately categorized as unpolluted with Zn, Pb, and Cu, and moderately polluted with Cd.

1 INTRODUCTION

The discharge of polluted effluents and waste into the water environment has increased dramatically and poses a serious threat to water organisms. Nutrients, heavy metals, and organic contaminants are accumulated in sediment to very high levels, turning the sediment into a large internal pollution source in aquatic systems. Sediment pollution is a major problem in aquatic ecosystem management and restoration. Nutrients, toxic chemicals, and toxin-forming microbes are found in much higher contents in sediments than in the overlying water column. Improvements in the quality of the overlying water and associated components of the aquatic ecosystem often cannot be achieved without some form of sediment remediation (Murphy et al., 1999). Removal of contaminated sediments from the water body is the most common approach for contaminated sediment remediation. Taihu Lake is the third largest freshwater lake in China with a surface area of 2338 km^2 and an average depth of about 2 m (Chen et al., 2003). It is situated in Changjiang (Yangtze) Delta, which is the most developed region in China (Qin et al., 2007). Eutrophication has caused a number of threats to water ecosystem in Taihu Lake, blooms of blue-green algae have been frequently occurring in warmer seasons (Pu et al., 1998). Sediment dredging has been carried out in Lake Taihu (Cao et al., 2007). Over 32,000,000 m^3 sediment has been dredged from Taihu Lake for the ecological restoration since 2008. Most of the dredged sediment has been seriously polluted by industrial wastewater and sewage. A large amount of dredged sediment was stored in dredge disposal zone around the lake without any treatment, which not only occupied a lot of land, but also caused secondary pollution.

In recent years, dredged material is increasingly regarded as a potential resource, useful for shoreline protection or for creation or restoration of habitats (in particular mudflats and salt marsh areas) in so-called beneficial use schemes (Ray, 2000; Yozzo et al., 2004; Bolam and Whomersley, 2005). So, the knowledge of the heavy-metal and nutrient contents of dredged sediments is necessary to next utilization.

The aim of the present study is to assess the heavy-metal and nutrient contamination of dredged sediments in Taihu Lake with special objectives:

1. to measure the heavy-metal and nutrient contents of dredged sediments to indicate the extent of contamination and
2. to determine the metal speciation of dredged sediments in order to assess the availability of heavy metals.

2 MATERIALS AND METHODS

2.1 Sample collection

Fifteen sediment samples were collected from disposal facility from Taihu Lake. The sediment samples were collected in November 2013 (rain season) and August 2014 (dry season). Then, 1 kg sediment was sampled from each point. Sampled sediments were stored in a clean PE sealing bag and then put in a foam box with ice packs at the bottom. The foam box, full of sediments, was carried back to the laboratory within 24 h. Sediments carried to laboratory were stored in a refrigerator or determined immediately. Sediment samples were pretreated as follows: first, the samples were air-dried; second, the stones, animal and plant residues, and other foreign matters were removed from it; third, they were ground and sifted through a 100-mesh sieve, and then finally were placed in a dry condition.

2.2 Testing items and methods

Sediment moisture content was determined by sediment weight lost before and after drying at 80°C to constant weights. Sediment pH was measured in a 1:10 (w/v) aqueous solution using a combined electrode (Thermo, USA). Organic Matter (OM) content in the sediment was determined by the potassium dichromate redox-volumetric method. Available nitrogen content in sediment was measured by an alkali diffusion solution. Available P is determined by $NaHCO_3$, extraction Mo–Sb colorimetric method, whereas available K is determined by atomic absorption spectrophotometry.

The concentrations of Cu, Zn, Pb, and Cd in the samples were analyzed. For metal analysis, total sediment digestion was performed in Teflon vessels following the classical open digestion procedures (SEPA, 2002). About 0.25 g air-dried sediment sample was put into a Teflon beaker and weighed, to it a mixture of concentrated HF–$HClO_4$–HNO_3 was added, then it was covered with a Teflon watch cover, and the sample was left at room temperature overnight. On the following day, the sample was heated to a temperature of about 200°C on a hot plate and kept under slight boiling state until the solid residue disappeared, and the solution turned into white or light yellow-greenish pasta-like material (Zhang et al., 2009). Then, 5 ml HNO_3 was added to completely dissolve it and make a final 50 ml solution with ultra-pure water. Concentrations of metals in solutions were determined using flame atomic absorption spectrometry (Ana-lytikjena AAS vario6, Germany) for Cu, Zn, and Pb and employing graphite furnace atomic absorption spectrometry for Cd.

2.3 Data analysis

The geoaccumulation index (I_{geo}) has been used since the late 1960s and has been widely employed in European trace metal studies. Originally used for bottom sediments (Müller, 1969), it has been successfully applied to the measurement of soil pollution (Loska and Wiechuła, 2003) and soil dust in city (Ji et al., 2008).

The I_{geo} enables the assessment of pollution by comparing current concentrations with preindustrial levels. It can be calculated by the following equation:

$$I_{geo} = \log_2\left(\frac{C_n}{1.5B_n}\right) \tag{1}$$

where C_n is the measured concentration of the examined metal n in the sediment, and B_n is the geochemical background concentration of the metal. The factor 1.5 is used because of possible variations in background values due to lithological variability.

3 RESULTS AND ANALYSES

3.1 *Nutrient characteristics of dredged sediment*

The physicochemical characteristics of the sediment collected from Taihu Lake were shown in Table 1.

The sediment moisture ranged from 22.36% to 29.24%. The moisture content is abundant enough to meet the needs of the plant growth. The sediment was alkaline in pH, with a range of 7.01–7.55, suitable for the growth of most plants. The OM has a spatial distribution, ranged from 0.7% to 6.1%. According to the soil nutrient classification standard in China (Table 2), the dredged sediment from Taihu Lake belongs to level 2. Available N content of sediment ranged from 35.99 to 139.95 ppm, which belongs to 4th soil nutrient level. The content of available P is lacking, ranging from 9.89 to 24.03 ppm, and is at 4th soil nutrient level. The rapidly available potassium content is between 12.24 and 27.98ppm, being at level 6 of soil nutrients. Therefore, rapidly available potassium in sediment of studied area is extremely deficient.

We can conclude that the OM and available phosphorus of dredged sediments from Taihu Lake are rich for plant growth, and the content of available nitrogen in dredged sediment from Taihu Lake is extremely deficient for the growth of plants, so additional sources of nitrogen and potassium should be added to the dredged sediment before its resource utilization.

3.2 *Analysis of heavy-metal content*

Heavy-metal concentrations are shown in Table 3. The coefficients of variations for Cu, Zn, Pb, and Cd were 25.58%, 54.20%, 37.33%, 23.82%, and 62.50%, respectively. Heavy-metal concentrations showed significant spatial variations in the study area. The comparative results (Table 3) showed that the concentrations of heavy metals studied in the Taihu Lake were clearer than those in the Dongting Lake, China; the Chaohu Lake, China; the Changjiang (Yangtze) River Estuary, China; and the Zhujiang (Peral) River Estuary, China.

Table 1. Physicochemical characteristics of sediment samples from Taihu Lake.

	Taihu Lake
pH	7.28 ± 0.27
Moisture content (%)	25.80 ± 3.44
Organic matter (%)	3.41 ± 2.69
Available N (ppm)	87.97 ± 51.98
Available P (ppm)	16.96 ± 7.07
Available K (ppm)	20.11 ± 7.87

Table 2. Soil nutrient classification standard in China.

Levels	Organic matter (%)	Available N (ppm)	Available P (ppm)	Available K (ppm)
1 (Very rich)	>4	>150	>40	>200
2 (Rich)	3–4	120–150	20–40	150–200
3 (Moderate)	2–3	90–120	10–20	100–150
4 (Lack)	1–2	60–90	5–10	50–100
5 (Very lack)	0.6–1	30–60	3–5	30–50
6 (Stunning lack)	<0.6	<30	<3	<30

3.3 Assessment based on geoaccumulation index

To understand current environmental status and heavy-metal pollution extent with respect to natural environment, I_{geo} is a common criterion to evaluate the heavy-metal pollution in sediments. It can be more correctly evaluated if complementary information based on metal baseline values is considered using the I_{geo} (Yu et al., 2008). In this study, we did not obtain the background concentrations of heavy metals in the sediments. Therefore, I_{geo} was calculated by using the soil element background values of Zhejiang Province. The results of I_{geo} values are shown in Table 5. Müller (1981) distinguished seven classes of I_{geo}, as shown in Table 6. From the classification criteria (Table 4), all the sediments collected from the study area could be approximately categorized as practically unpolluted with Zn, Pb, and Zn (mean I_{geo} < 0 for each heavy metal) and moderately polluted with Cd (1 < mean I_{geo} < 2 for both heavy metals). On the basis of the mean values of I_{geo}, the degree of heavy-metal pollution in dredged sediments was found to be low.

Table 3. Mean concentrations of trace metals from the Taihu Lake sediments and published mean sediment values in sediments.

	Cu	Zn	Pb	Cd
	(mg/kg)	(mg/kg)	(mg/kg)	(mg/kg)
Maximum value	17.02	49.20	16.44	0.22
Minimum value	11.76	21.22	9.41	0.08
Mean value	15.17	36.68	14.33	0.16
Standard deviation	3.88	19.89	5.35	0.1

Table 4. Reference values of heavy metals.

	Testing items			
Regions	Cu	Pb	Zn	Cd
Averages of the bottom sediment of the whole Taihu Lake	41.38	41.00	103	0.79
Background values of water sediments of Yangtze River	21.50	21.40	73.6	0.148
Averages of the bottom sediment of Poyang Lake	27.56	12.50	71.1	0.750
Averages of the bottom sediment of the whole Dongting Lake	51.3	52.8	140	2.7
Background levels of soil, China	35.0	35.0	100	0.2
Threshold values for severely polluted soil, China	400	500	500	1.0

Table 5. I_{geo} and pollution levels of Cu, Pb, Zn, and Cd in sediments from Taihu Lake.

		Taihu Lake
Cu	Max	−0.66/0
	Min	−1.46/0
	Mean	−0.98/0
Pb	Max	−0.33/0
	Min	−1.13/0
	Mean	−0.68/0
Zn	Max	−2.08/0
	Min	−3.54/0
	Mean	−2.61/0
Cd	Max	1.01/2
	Min	−0.51/0
	Mean	0.49/1

Table 6. Contamination degree corresponding to index of geoaccumulation.

I_{geo}	0	0~1	1~2	2~3	3~4	4~5	>5
Grades	0	1	2	3	4	5	6
Pollution levels	Clear	Clearly moderate	Moderate	Moderately strong	Strong	Strong, extremely strong	Extremely strong

3.4 Conclusions

1. The pH and moisture content of dredged sediment from the Taihu Lake could meet the requirements for the growth of most plants.
2. OM content in the Taihu Lake can satisfy the landscape use. Available N and K are very low in most of the research areas, so additional sources of N and K should be added to ensure the normal growth of plants for phytoremediation. Apart from individual sample points, available P could satisfy the growth of most plants.
3. Although the dredged sediment from the Taihu Lake suffered from various degrees of Cu, Pb, and Zn pollution, it can meet the requirement of resource utilization. However, Cd pollution is generally serious and has a certain concentration effect. Therefore, plantation should be made for phytoremediation, which will lead to the reduction of Cd content in dredged sediment.

ACKNOWLEDGMENT

This work is supported by special fund for basic scientific research business of central public research institutes TKS 130103.

REFERENCES

Bolam S.G. & Whomersley P. 2005. Development of macrofaunal communities on dredged material used for mudflat enhancement: a comparison of three beneficial use schemes after one year. Marine Pollution Bulletin 50(1): 40–47.

Cao X.Y., Song C.L., Li Q.M., et al. 2007. Dredging effects on P status and phytoplankton density and composition during winter and spring in Lake Taihu, China. Hydrobiologia 581(1): 287–295.

Chen Y.W., Fan C.X., Teubner K., et al. 2003. Changes of nutrients and phytoplankton chlorphyll a in a large shallow lake, Taihu, China: An 8-year investigation. Hydrobiologia 506/509: 273–279.

Ji Y.Q., Feng Y.C., Wu J.H. et al. 2008. Using Geoaccumulation index to study source profiles of soil dust in China. Journal of Environmental Science 20(5): 571–578.

Loska K., Wiechuła D. 2003. Application of principal component analysis for the estimation of source of heavy metal pollution in surface sediments from the Rybnik Reservoir. Chemosphere 51(8): 723–733.

Müller G. 1969. Index of geoaccumulation in sediments of the Rhine River. Geojournal 2(1): 108–118.

Müller G. 1981. Die Schwermmetallbelastung der sedimente des Neckars und seiner Nebenflusse: eine Bestandsaufnahme. Chemical Zeitung 105(2): 157–164.

Murphy T.P., Lawson A., Kumagai M., et al. 1999. Review of emerging issues in sediment treatment. Aquatic Ecosystem Health & Management 2(2): 419–434.

Pu P.M., Hu W.P., Yan J.S., Wang G.X., Hu C.H. 1998. A physicoecological engineering experiment for water treatment in a hypertrophic lake in China. Ecological Engineering 10(2): 179–190.

Qin B.Q., Xu P.Z., Wu Q.L., et al. 2007. Environmental issues of Lake Taihu China. Hydrobiologia 581(1): 3–14.

Ray G.L. 2000. Infaunal assemblages on constructed intertidal mudflats at Jonesport, Maine (USA). Mrine Pollution Bulletin 40(12): 1186–1200.

SEPA (State Environmental Protection Administration of China). Fourth ed. 2002. Water and waste water analysis. Beijing: Chinese Environmental Science Publish House.

Yozzo D.J., Willber P., Will R.J. 2004. Beneficial use of dredged material for habitat creation, ehancment, and restoration in New York-New Jersey Harbor. *Journal of Environment Management* 73(1): 39–52.

Yu R.L., Yuan X., Zhao Y.H. et al. 2008. Heavy metal pollution in intertidal sediments from Quanzhou Bay, China. *Journal of Environmental Science* 20(6): 664–669.

Hydraulic Engineering IV – Xie (Ed.)
© 2016 Taylor & Francis Group, London, ISBN 978-1-138-02948-4

Synthesis and properties of MnO$_2$/polypyrrole hollow urchins

Hong Shi, Yang Tian, Feng Chen, Shi Lin Zeng, Jiaying Zuo, Decai Li, Lingfeng Li & Xiang Hong Peng
Key Laboratory of Optoelectronic Chemical Materials and Devices, Ministry of Education, School of Chemical and Environmental Engineering, Jianghan University, Wuhan, China

ABTRACT: MnO$_2$/polypyrrole hollow urchins were synthesized using hollow urchins MnO$_2$ as template, in which it was synthesized by a hydrothermal reaction in the 0.3 M KMnO$_4$ and 0.15 M (NH$_4$)$_2$SO$_4$ solution. The structures and the morphologies of MnO$_2$ and MnO$_2$/Ppy hollow urchins were characterized by Fourier transform infrared spectroscopy, SEM and cyclic voltammogram. The result showed that MnO$_2$ hollow urchins were formed by the nanorods which were 100–300 nm in diameters, having 1–3 μm hollow diameters. The specific capacitance value of MnO$_2$/Ppy hollow urchin was 195.3 F/g at a scan rate of 10 mV/s. Such hollow urchin possessing a large surface may provide a novel electrode material for electrochemical energy storage.

1 INTRODUCTION

Developing efficient energy storage and conversion have drawn increasing attention. The Supercapacitors (SCs) have been considered as excellent potential candidates for energy storage because of their higher power density, longer cycle stability, and greater energy density (Yan, 2014; Wang, 2014). There are increasing efforts of the electrode materials of SCs, such as carbonaceous materials, metal oxides or hydroxides, and conducting polymers (Yu, 2013; Ellis, 2014; Li, 2014). Among the various supercapacitive electrodes materials, MnO$_2$ is one of the most promising materials for SCs because of its low cost, large theoretical specific capacitance, and environment-friendly property. However, the poor conductivity of MnO$_2$ prevents its wide energy storage applications. There remains a major challenge to synthesize the orderly MnO$_2$ hybrid electrodes with high-weight fraction and large specific surface area (Yu, 2013; Yu, 2009). MnO$_2$ hollow urchins have a large surface area as urchins were formed by the MnO$_2$ nanorods (Xu, 2007). Conducting polymers, including polypyrrole (Ppy), polythiophene, polyaniline, and their derivatives have shown wide applications in the field of the electrode materials for SCs, because of their special optical, electrical, and biocompatible properties (Wang, 2014). As reported, the synergistic effect produced by a combination of multiple materials would overcome the limitation of single materials and improve specific capacitance, such as various MnO$_2$/conductive matrix hybrid materials (Zhang, 2015; Yao, 2013; Wang, 2014; Grote, 2014). In this paper, we report synthesis of MnO$_2$/Ppy hollow urchins using MnO$_2$ hollow urchins as template, in which MnO$_2$ hollow urchins were synthesized by a hydrothermal reaction. Such a combination of MnO$_2$ and Ppy having a large surface may help the electrode materials for SCs.

2 EXPERIMENTAL DETAILS

2.1 Materials

Pyrrole monomer (99%) was purchased from Aladdin Company. The other agents at analytical reagent grade were obtained from Sinopharm Chemical Reagent Co., Ltd., Shanghai.

Preparation of multilayer MnO₂/Ppy hollow urchins: The MnO_2 products were synthesized by a hydrothermal reaction with a 0.3 M $KMnO_4$ solution and a 0.15 M $(NH_4)_2SO_4$ solution for 24 h at 90°C. The products were filtered, washed, and dried at 60°C for 6 h to obtain the MnO_2 powder. An amount of 0.05 g of pyrrole monomer was dissolved in deionizer water and then added to the above MnO_2 powder. The final products were filtered, washed, and dried at 60°C for 12 h to obtain the MnO_2/Ppy hollow urchins.

Characterization: Fourier Transform Infrared Spectroscopy (FTIR) of MnO_2 and MnO_2/Ppy hollow urchin powder was recorded on a TENSOR 27 spectrometer (Bruker Optics, Germany) in a range from 4000 to 400 cm^{-1}. The morphologies of the microencapsules were characterized on a scanning electron microscope (SU-8000, Hitachi, Japan).

2.2 Electrochemical measurement

The working electrodes of electrochemical capacitors were formed by the mixing of prepared powder with 15 wt% acetylene black and 5 wt% poly-(tetrafluoroethylene) (PTFE) binder of the total electrode mass. A small amount of distilled water was then added to those mixtures to make them more homogeneous. The mixtures were pressed onto nickel foam current collectors (1.0 cm^2) to fabricate the electrodes. All electrochemical measurements were carried in a three-electrode experimental setup. Platinum foil with the same area as the working electrode and a Saturated Calomel Electrode (SCE) were used as the counter and reference electrodes, respectively. All the electrochemical measurements were carried out in 1 M Na_2SO_4 aqueous electrolyte by using a CHI 660E electrochemical workstation (CHI Instruments).

3 RESULTS AND DISCUSSION

The FTIR spectra of MnO_2 and MnO_2/Ppy hollow urchins are shown in Figure 1. The peaks at 1565 cm^{-1} and 1415 cm^{-1} belong to the backbone stretching vibrations of $C=C$ and $C–C$ in pyrrole ring. The peaks at 1040 cm^{-1} and 919 cm^{-1} are assigned to the in-plane deformations of $=C–H$ and $N–H$, respectively [9]. As a result, the C–N stretching vibration at 1190 cm^{-1} and the C–H out-of-plane vibration at 919 cm^{-1} suggest that Ppy is in a doping state. The above results confirm the existence of interfacial interaction between Ppy and MnO_2.

The SEM images of MnO_2 and MnO_2/Ppy hollow urchins are shown in Figure 2. The MnO_2 hollow urchins exhibit a hollow structure with 0.5–1 μm diameters. The hollow urchins are formed by densely aligned nanorods which have 100–200 nm diameters. The samples of MnO_2/Ppy also show the hollow urchins structure, which is built up with the MnO_2/Ppy

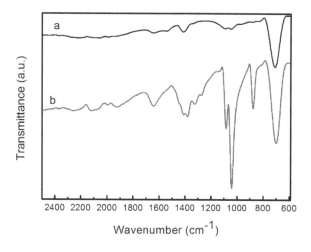

Figure 1.   FTIR spectra of (a) Ppy and (b) Ppy/MnO_2.

Figure 2.   SEM of (a) MnO$_2$ hollow urchins, (b) MnO$_2$/Ppy hollow urchins.

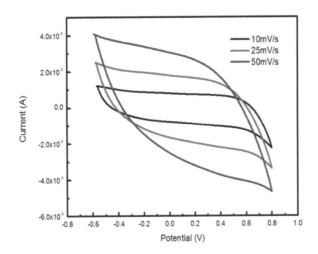

Figure 3.   CV of MnO$_2$/Ppy hollow urchins.

nanorods, having 100–300 nm diameters. As reported, the most of the Ppy morphology is like cauliflower; the morphology of Ppy nanorods in this work indicates that the pyrrole mono-mer was polymerized on the surface of MnO$_2$ and induced by the MnO$_2$ nanorods because of the interaction between MnO$_2$ and Ppy. The MnO$_2$/Ppy hollow urchins have higher specific capacitance, which is 195.3 F/g.

The Cyclic Voltammograms (CVs) for MnO$_2$/Ppy hollow urchin electrodes are shown in Figure 3. The specific capacitances of MnO$_2$/Ppy hollow urchins are 195.3, 156.8, and 100.7 F/g at scan rates of 10, 20, and 50 mV/s, respectively. With the increase of scan rate, the specific capacitances of MnO$_2$/Ppy hollow urchins gradually decrease. In general, the redox reactions depend on the insertion and extraction rate of protons or alkali cations from the electrolyte. At a lower scan rate, the working ions can reach almost all active areas, resulting in high effective interaction between the ions and the electrode (Yao, 2013; Wang, 2014).

4   CONCLUSIONS

MnO$_2$/Ppy hollow urchins were prepared by using MnO$_2$ hollow urchins as template, which have 100–300 nm diameters and 0.5–1 μm hollow diameter. The specific capacitance value was 195.3 F/g at a scan rate of 10 mV/s. Such hollow urchins, having a large surface area, may be a novel electrode material for SCs.

ACKNOWLEDGMENTS

This work was supported by the Opening Project of Key Laboratory of Optoelectronic Chemical Materials and Devices (Jianghan University), Ministry of Education (JDGD-201602) the Foundation of Science and Technology Bureau of Wuhan (2014010101010016).

REFERENCES

Ellis, B.L.; Knauth, P.; Djenizian, T. Three-Dimensional Self-Supported Metal Oxides for Advanced Energy Storage. Adv. Mater. 2014, 26, 3368–3397.

Grote, F.; Lein, Y. A complete three-dimensionally nanostructured asymmetric supercapacitor with high operating voltage window based on Ppy and MnO_2. Nano Energy. 2014, 10, 63–70.

Li, P.; Yang Y.; Cao, A.; Core-Double-Shell, Carbon Nanotube@ Polypyrrole@ MnO_2Sponge as Freestanding, Compressible Supercapacitor Electrode. ACS Appl. Mater. Interfaces 2014, 6, 5228–5234.

Wang, J.; Yang, Y.; Huang, Z.; Kang, F. MnO_2/polypyrrole nanotubular composites: reactive template synthesis, characterization and application as superior electrode materials for high-performance supercapacitors. Electrochimica Acta, 2014, 130, 642–649.

Wang, K.; Wu, H.; Meng, Y.; Wei, Z. Conducting Polymer Nanowire Arrays for High Performance Supercapacitors. small 2014, 10, 14–31.

Xu, M.; Kong, L.; Zhou, W.; Li, H. Hydrothermal Synthesis and Pseudocapacitance Properties of r- MnO_2 Hollow Spheres and Hollow Urchins. J. Phys. Chem. C 2007, 111, 19141–19147.

Yan, J.; Wang, Q.; Wei, T.; Fan, Z.; Recent Advances in Design and Fabrication of Electrochemical Supercapacitors with High Energy Densities. Adv. Energy Mater. 2014, 4, 1300816 (1–43).

Yao, W.; Zhou, H.; Lu, Y. Synthesis and property of novel MnO_2@polypyrrole coaxial nanotubes as electrode material for supercapacitors. Journal of Power Sources, 2013,241, 359–366.

Yu, P.; Zhang, X.; Wang, D.; Wang, L.; Ma, Y. Shape-Controlled Synthesis of 3D Hierarchical MnO_2 Nanostructures for Electrochemical Supercapacitors. Crystal Growth & Design, 2009, 9, 528–533.

Yu, Z.; Duong, B.; Abbitt, D.; Thomas, J. Highly Ordered MnO_2 Nanopillars for Enhanced Supercapacitor Performance. Adv. Mater. 2013, 25, 3302–3306.

Zhang, Q.; Liu, Z.; Wang, K. Zhai, J. Organic/Inorganic Hybrid Nanochannels Based on Polypyrrole-Embedded Alumina Nanopore Arrays: pH- and Light-Modulated Ion Transport. Adv. Funct. Mater. 2015, 25, 2091–2098.

Hydraulic Engineering IV – Xie (Ed.)
© 2016 Taylor & Francis Group, London, ISBN 978-1-138-02948-4

Comprehensive analysis of dust removal efficiency of gas–water nozzle

Ming Wang & Zhongan Jiang
University of Science and Technology Beijing, Beijing, China

ABSTRACT: To improve the dust removal efficiency of gas–water nozzle spray in coal mine workplaces of high-concentration dust, gas–water nozzle atomization characteristic parameters have been realized through experiments, and the changing rule between droplet average diameter and gas/water flow was concluded. This paper studied the dust removal process of gas–water spraying in comprehensive tunneling face, establishing the corresponding mathematical model, deducing the relation formula for gas–water nozzle spray efficiency, drawing the curve of dust removal efficiency by using MatLab. The research shows that, on the one hand, when the water flow remains constant, dust removal efficiency increased with the increase of gas flow, on the other hand, when the gas flow remains constant, dust removal efficiency first increased and then decreased as the water flow increased. There exists the best gas–water flow ratio to maximize the dust removal efficiency. The larger the dust size, the easier the dust could settle. Based on the dust size distribution and removal requirement of dust in working face, both a better dust removal effect and economic benefit can be achieved by choosing the best gas/water flow ratio, with reference to the relationship curve.

1 INTRODUCTION

Plenty of dust will be generated in the process of operation in mining face of coal mines, and wet dust collector is the most economical and convenient measure to handle it (Zuo, 2014). But conventional pressure water nozzles often require high water pressure and large water volume. Moreover, because the particle size of the droplet is large, the dust removal efficiency is low (Li, 2011). Gas–water nozzle atomizing aspirating is an emerging technology in China, and both domestic and foreign research show that it is much more efficient than conventional spray dust removal technology, regarding the reduction of breathing dust concentration (Sarkar, 2007; Hou, 2014). But the truth is, with relatively few theoretical studies of gas–water nozzle spray dust removal, this technology is still to mature in practice. Thus, most of the adjustment of gas and water flow is merely experiential, which not only wastes water but also lowers the expected dust-controlling effect. Based on the experimental research, combining with theoretical analysis of atomization mechanism, this paper studied the atomization characteristics of the gas–water nozzle and analyzed the factors that influence the dust removal efficiency of gas–water nozzle spray. Through this, it concluded the efficiency curve among gas–water spray and air flow, water flow and gas flow ratio, which provides a theoretical guidance to the practical application of the gas–water nozzle.

2 GAS–WATER NOZZLE ATOMIZATION CHARACTERISTIC RESEARCH

2.1 *The structure and principle of the gas–water nozzle*

Gas–water nozzle comprises four main parts, which are water intake port, air intake port, air–water mixing chamber and the export of spray, as shown in Figure 1. Its principle is as follows: gas and water with certain pressure come in from water inlet and air inlet, respectively; then

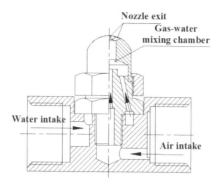

Figure 1.   Schematic diagram of gas–water nozzle.

the water flow is broken into wire or liquid line with a large number of tiny air bubbles under the effect of high-speed air flow (Cao, 2013; Li, 2006). After that, they form stable-bubbles mixed flow in the mixing chamber. When the mixture comes out the nozzle with high speed, due to the volume expansion of the mixture, the agitation of the fluid and the involvement of the surrounding air, water is atomized into many small drops (Liu, 2001; Wang, 2013).

2.2   Analysis of gas–water nozzle's flow characteristics

In practice, dust-controlling effect of the gas-water nozzle is associated with the parameters, namely dust particle diameter, jet parameters, distance and covering angle (Daviault, 2012). These parameters are not only associated with the mining chamber and nozzle structure, but also depend on gas flow rate, water flow and gas flow ratio of the gas–water nozzle. Therefore, to measure them, the nozzle pressure gauge and flow meter are installed with high-pressure pipes at the end of the air and water intake port. During the research of the gas–water nozzle flow characteristics, air compressor and QL-380 A type cleaning machine were used to provide air and water separately, and the gas and water flow of nozzle were adjusted by changing the air and hydraulic pressure. As a result, the relationship between water flow and the gas flow ratio is shown in Figure 2.

It can be concluded from Figure 2 that no matter how gas or water pressure changes, water flow always keeps the exponential relation with gas flow ratio, as shown in the fitting formula (1):

$$Q_l = 5.83e^{-0.03R} \times 10^{-5} \tag{1}$$

Among them, $R = Q_g / Q_l$ represents the nozzle gas–water volume flow ratio; Q_g and Q_l represent the nozzle flow of gas and water flow (m³/s), respectively.

2.3   Analysis of gas and water nozzle droplet size

For the gas–water nozzles that have similar structure and principle, they maintain the same similarity criterion. The related experience formula for the average diameter of in-mixed type gas–water atomization nozzle droplets gained from references (Barroso, 2014; Hou, 2010; Nguyen, 1998; Raj, 2008) is shown below.

$$D_w = \frac{585\sqrt{\sigma}}{V_R\sqrt{\rho_l}} + 597\left(\frac{\mu_l}{\sqrt{\sigma\rho_l}}\right)^{0.45}\left(1000\frac{Q_l}{Q_g}\right)^{1.5} \tag{2}$$

where D_w is the average droplet diameter type (μm); V_R is the relative velocity of gas to liquid (m/s); σ is the surface tension coefficient of liquid (dyn/cm); μ_l is the liquid viscosity coefficient (dyn·s/cm²); ρ_a is the density of water (g/cm³).

In formula (2), $V_R = \sqrt{(V_g \sin \alpha)^2 + (V_g \cos \alpha - V_l)^2}$, where V_g and V_l represent the velocity of air and water inside the nozzle (m/s), respectively; $\alpha = 30°$ represents the gas and water tunnels angle. Putting all these parameters in formula (2) gives

$$D_w = \frac{0.017}{\sqrt{(1.60Q_g)^2 + (2.7Q_g - 5.7Q_l)^2}} + 0.28\left(\frac{Q_l}{Q_g} \times 10^3\right)^{1.5} \qquad (3)$$

The trend curve of the average particle size of droplets changing with gas and water flow is plotted by using the MatLab software, as shown in Figure 3. From Figure 3, it is concluded that when the water flow remains constant, droplet average particle size decreases with the increase in pressure; when the gas flow is constant, the greater the water flow, the greater the droplet's particle size. When the water flow is less than 1×10^{-5} m³/s, and gas flow is greater than 50×10^{-5} m³/s, the average particle size of droplets is less than 10 μm.

3 THE MATHEMATICAL MODEL OF THE GAS–WATER NOZZLE SPRAY FOR DUST REMOVAL

3.1 *Assuming conditions*

This paper takes comprehensive driving face in coal mines, as example, to conduct analysis; an external spray module is installed in the rocker arm of the heading machine, which sprays

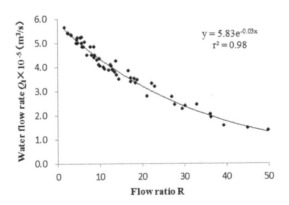

Figure 2. Relationship curve between water flow and gas/water flow ratio.

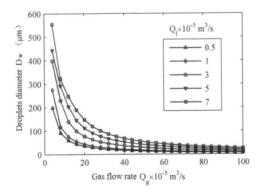

Figure 3. Relationship curve between dust particle sizes and gas flow under different water flows.

223

mist to tunnel face when the machine excavates coal, and it forms mist flow column near heading face after injection becomes stable. Assuming that the distribution uniformity of spray sprinkling is good, and multiple-jet diffusion angles can cover the whole cross-section of roadway, then water mist and dust particles could collide and precipitate within the spraying range of water mist. Within this range, the velocity of water mist is much faster than that of dust-containing air flow, so its relative velocity can be approximately regarded as the velocity of droplets (Ma, 2005).

3.2 Establishment of mathematical model

Single water droplet's ability to capture dust particles is one important parameter to reflect the efficiency of dust removal by spraying. As shown in Figure 4, air passes over a water droplet with a velocity $\mathbf{U}_r$ relative to the droplet, the streamlines of air bend around the droplet. However, the inertia of dust particles causes them to cross those streamlines. Particles that lie around the center line of motion will collide with the droplet and be captured.

We can conceive a flow tube of diameter y, within which all particles are captured, whereas particles that are further from the tube center line will be diverted around the droplet. So the capture efficiency of a single droplet E can be defined as the ratio of the cross-sectional areas of the capture tube to the facing area of the droplet

$$E = \frac{y^2}{D_w^2} \tag{4}$$

Assuming that dust is evenly distributed in the air, and per cubic meter of it contains n particles, then the capture rate of the particles by one droplet is

$$N = n\mathbf{U}_r \pi \frac{y^2}{4} = En\mathbf{U}_r \pi \frac{D_w^2}{4} \tag{5}$$

To maintain consistency with the definition of dust concentration that we used here (particles/m³), it is better to restate the later expression in terms of particles collected per cubic meter of air rather than particles captured per second. We can do this by dividing by the air flow rate Q (m³/s). Then the capture rate of one droplet (dn/dt = rate of change of dust concentration) becomes

$$-\frac{dn}{dt} = En\mathbf{U}_r \pi \frac{D_w^2}{4Q} \tag{6}$$

Now if water is dispersed in the spray at a volume flow rate of W (m³/s), then the rate at which droplets are formed and pass through the spray is $W/(\pi D_w^3/6)$. Multiplying by the

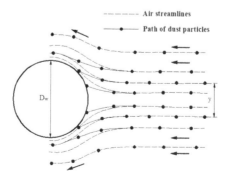

Figure 4. Inertia-effect diagram of single droplet to dust particle.

particle capture for one particle, equation (6) gives the total rate at which particles are captured per cubic meter of air:

$$-\frac{dn}{dt} = EnU_r\pi\frac{D_w^2}{4Q}\frac{W}{\pi D_w^3/6} = \frac{3}{2}EnU_r\frac{W}{QD_w}$$ (7)

Now let us consider Figure 5 where dust particles and air pass each other effectively in counter flow with a relative velocity of U_r, which means that they move through a separation distance dx in time dt.

$$U_r = \frac{dx}{dt}$$ (8)

During that time, the dust concentration changes from n to $n-dn$. The change rate of dust concentration is

$$-\frac{dn}{dt} = -\frac{dn}{dx}\frac{dx}{dt} = -\frac{dn}{dx}U_r$$ (9)

Combining with equation (7) gives

$$dn = -\frac{3}{2}En\frac{W}{QD_w}dx$$ (10)

Integrating with the complete effective length of the spray L (distance moved by particles plus distance moved by droplets in the x direction) gives the total number of particles removed between the inlet concentration, n_1, and outlet concentration, n_2 (particles/m³).

$$\int_{n_1}^{n_2}\frac{dn}{n} = -\frac{3}{2}En\frac{W}{QD_w}\int_0^L dx$$ (11)

Solving equation (11) gives

$$\frac{n_2}{n_1} = \exp\left(-\frac{3EWL}{2QD_w}\right)$$ (12)

The dust removal efficiency of spray is given by

$$\eta = \frac{n_1 - n_2}{n_1} = 1 - \frac{n_2}{n_1} = 1 - \exp\left(-\frac{3EWL}{2QD_w}\right)$$ (13)

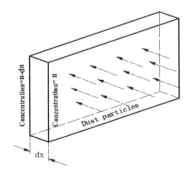

Figure 5. Tiny body.

Only considering the inertial effect, Wong et al. put forward dust reducing efficiency of a single droplet inertia collision (Zhang, 1987):

$$E = \left(\frac{K}{K + 0.7} \right)^2 \tag{14}$$

The dimensionless parameter K is

$$K = \frac{U_r \rho_p D_P^2}{9 \mu_g D_w} \tag{15}$$

where ρ_p is the particle density (kg/m³); D_p is the particle diameter (m); μ_g is the kinematic viscosity of air (Pa.s).

The relative velocity of droplets and dust particles in driving faces approximate to half of the nozzle spray exit velocity:

$$U_r = \frac{Q_l + Q_g}{2 A_0} \tag{16}$$

Solving equation (13) – (16) gives

$$\eta = 1 - exp\left[-\frac{3WL}{2AU_g D_w} \times \left(\frac{(Q_l + Q_g)\rho_p D_P^2}{(Q_l + Q_g)\rho_p D_P^2 + 12.6 \mu_g D_w A_0} \right)^2 \right] \tag{17}$$

Take a coal mine as example. The wind velocity of comprehensive driving face U_g is 1.5 m/s; sectional area of roadway A is 15 m², dust particle density ρ_p = 600 kg/m³, kinematic viscosity of the air $\mu_g = 1.8 \times 10^{-5}$ Pa.s, the spray device has six nozzle in total, then W = 6 Q_l. The nozzle exit area A0 = 3.14 × 10⁻⁶ m², putting all these parameters in equation (17) gives

$$\eta = 1 - exp\left[-4.05 \frac{Q_l L}{D_w} \times \left(1 + 0.12 \frac{D_w}{(Q_l + Q_g)D_P^2} \times 10^{-12} \right)^{-2} \right] \tag{18}$$

4 ANALYSIS OF DUST REMOVAL EFFICIENCY OF GAS–WATER NOZZLE

The main factors that influence the gas–water spray application in tunneling working faces include gas flow, water flow, gas flow ratio, particle size distribution and other factors. Combining with formula (18), MatLab software is used to simulate the dust removing efficiency curve with various affecting factors.

1. Relationship among dust reducing efficiency, gas flow and water flow of gas–water nozzle.

Figures 6 and 7 show how dust removing efficiency changes under different gas or water flow. As shown in the figures, (1) when the water flow is fixed, the greater the gas flow, the higher the dust reducing efficiency of gas–water spray device will be. Especially, when the gas flow rate is less than 150 × 10⁻⁵ m³/s, the dust removing efficiency significantly increases with the increase of gas flow; (2) when the gas flow is fixed, the dust reducing efficiency first increases and then decreases with the increase of water flow. So as the gas flow is given, there is a best water flow to make the dust reducing efficiency highest. The reason is that the dust reducing efficiency is not only related to particle diameter but also the number of droplets. The more the water droplets, the better the dust reducing efficiency is. Meanwhile, the particle diameter gets bigger as water flow gets bigger, thus affecting dust reducing efficiency.

226

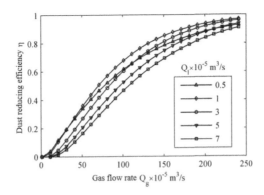

Figure 6. Dust reducing efficiency changing relationship with gas flow under different water flows.

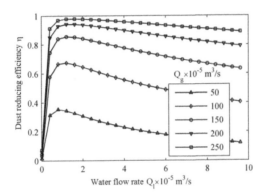

Figure 7. Dust reducing efficiency changing relationship with water flow under different gas flows.

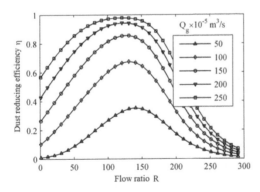

Figure 8. Dust reducing efficiency changing relationship with gas/water flow ratio under different gas flows.

2. Relationship between dust reducing efficiency and gas flow ratio of gas–water nozzle.

As shown in Figure 8, relationship between dust reducing efficiency and gas flow ratio under different gas flows is simulated by the MatLab. When the gas flow remains fixed, the dust reducing efficiency first increases and then decreases with the increase in gas flow ratio. So there exists a best gas flow ratio that makes the dust reducing efficiency highest. When the gas flow is 200×10^{-5} m³/s, the best gas flow ratio is 125, and under this condition, the dust reducing efficiency is above 90%.

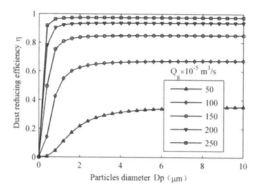

Figure 9.   With a water flow of 1×10^{-5} m³/s, the relationship between dust reducing efficiency and the particle sizes of dust under different gas flows.

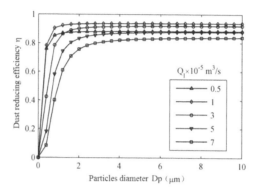

Figure 10.   With a gas flow of 1×10^{-5} m³/s, the relationship between dust reducing efficiency and the particle sizes of dust under different water flows.

3. Relationship between dust reducing efficiency and dust particle size.

Figure 9 plots the relationship between dust reducing efficiency and dust particle size under different gas flow, when the water flow is 1×10^{-5} m³/s. As indicated in Figure 9: (1) dust reducing efficiency stably increases as dust particle size increases and then reaches its maximum; (2) when water flow is constant, the greater the gas flow, the higher the dust reducing efficiency will be. Moreover, when dust efficiency tends to be stable, the sizes of the particles that can be captured tend to be smaller. When the gas flow is 150×10^{-5} m³/s, the dust removal efficiency is higher than 80% for the dust particles above 2.5 μm; whereas for the dust particles size is under 2.5 μm, the dust reducing efficiency falls sharply. When the gas flow is 200×10^{-5} m³/s for the dust whose particle size is above 2 μm, the reducing efficiency is higher than 90%; however, if the dust reducing efficiency is expected to reach more than 90%, for dust particles of 1 μm, the gas flow must be greater than 250×10^{-5} m³/s.

Figure 10 describes the dust reducing efficiency changes with a dust particle diameter at different water flows, when gas flow is 200×10^{-5} m³/s. From the figure, it can be seen that (1) when the gas flow is big enough, and the water flow is greater than 0.5×10^{-5} m³/s, the dust reducing efficiency is higher than 80% for the dust whose size is above 2.5 μm; (2) when the gas flow is fixed, there is a best water flow that makes the dust reducing efficiency maximum.

5   CONCLUSIONS

1. When gas–water nozzle spray is applied to reduce dust, the efficiency mainly depends on its gas flow, water flow, air water flow ratio and the coal dust particle size distribution.

2. The lager the gas flow and the smaller the particle diameter, the better it is for the settlement of dust with smaller particle sizes. When the water flow and the particle diameter get larger, it is not easy for dust settlement. But the lager water flow increases the moisture content between the working space, as well as the collision probability between dust particles and droplets, through which it improves the dust removing efficiency.
3. For any given gas flow, there is a best water flow that enables the largest dust removing efficiency of gas nozzle spray.
4. According to the dust particle size distribution in the working surface, dust removing efficiency requirements, proper air flow and water flow can be chosen to get the best effect of dust and economic benefits with reference to the corresponding curve.

REFERENCES

Barroso, J. et al. 2014. Analysis and prediction of the spray produced by an internal mixing chamber twin-fluid nozzle. Fuel Processing Technology 128: 1–9.

Cao, J.M. et al. 2013. Study on air assistant to improve quality of droplet atomization. Journal of Experiments in Fluid Mechanics 27(1): 56–60.

Daviault, S.G. et al. 2012. Atomization performance of petroleum coke and coal water slurries from a twin fluid atomizer. Fuel 98: 183–193.

Hou, L.Y. & Hou, X.C. 2010. Nozzle Technical Manual. Bei Jing: China Petrochemical Press.

Hou, T.Y. et al. 2014. Study on droplet breakup mechanism and dust-fall efficiency of mine wind-water atomizer. Mining Machinery 42(7): 132–135.

Li, P. & Zhang, W. 2006. Particle diameter investigation on internal mixing air-liquid atomizer. Small Internal Combustion Engine and Motorcycle 35(4): 21–24.

Li, Y.Q. et al. 2011. New Progress on Coal Mine Dust in Recent Ten Years. Procedia Engineering 26: 738–743.

Liu, L.S. et al. 2001. Experimental studies on the spray characteristics of effervescent atomizers. Journal of Combustion Science and Technology 7(1): 62–66.

Ma, S.P. & Kou, Z.M. 2005. Study on mechanism of reducing dust by spray. Journal of China Coal Society 30(3): 297–300.

Nguyen, D.A. & Rhodes, M.J. 1998. Producing fine drops of water by twin-fluid atomization. Powder Technology 99(3): 285–292.

Raj, B.M. et al. 2008. Comprehensive analysis for prediction of dust removal efficiency using twin-fluid atomization in a spray scrubber. Separation and Purification Technology 63(02): 269–277.

Sarkar, S. et al. 2007. Modeling of removal of sulfur dioxide from flue gases in a horizontal cocurrent gas-liquid scrubber. Chemical Engineering Journal 131(1–3): 263–271.

Wang, Y.J. & Zhang, T.L. 2013. Application of air-water spray humidification dust elimination technology in painting shop. Paint & Coatings Industry 43(3): 70–72.

Zhang, G.Q. 1987. Aerosol Mechanics. Bei Jing: China Environmental Science Press.

Zuo, G.L. et al. 2014. Study on optimization of internal and external spray system on shearer. Coal Technology 11: 226–228.

Hydraulic Engineering IV – Xie (Ed.)
© *2016 Taylor & Francis Group, London, ISBN 978-1-138-02948-4*

Study and practice on dust control technology in fully mechanized excavation face

Zhong-an Jiang, Yi-kun Zhang & Ming Wang
*School of Civil and Environmental Engineering, University of Science and Technology—Beijing,
Beijing, China*

ABSTRACT: Due to deficiencies in the existing methods for removal of dust from coal mines at a fully mechanized excavation face, a more effective method is proposed in this study. It applies the swirl effect of the ventilator that is attached to the wall in the way of far-pressing-near-absorption ventilation system to gather high concentration dust to face and then combines with gas water spray to prevent and control dust. Taking 21,007 fully mechanized excavation face of coal mine as the prototype and according to actual conditions under the mine, GAM-BIT software is used to establish a roadway geometry. Fluent software is used to simulate the roadway and to get the results of dust distribution with different ventilation methods. Then the simulation results are contrasted with the experiment results to get the high dust concentration zones. Based on the above results, self-actuated wet scrubber, wall attaching chimney and original gas water spray are used to prevent dust. The data from the site measurement show that dust suppression efficiency of full dust and respirable dust reaches 87%.

1 INTRODUCTION

As one of the major hazards of coal mine, coal mine dust includes rock dust and coal dust, which can cause pneumoconiosis and coal dust explosion. Coal mining mechanization and mining strength increase quickly, leading to a sharp increase of producing dust capacity of underground work place, which seriously threatens the safety of coal mine production and miners' health. Consequently, the workers can easily suffer from pneumoconiosis disease for working in a long-term dusty environment. At the same time, when the dust reaches a certain concentration, a violent explosion can occur with a heat source, which will have extremely painful consequences. In addition, the loss will be more substantial if the coal dust explodes with gas.

To strengthen the control of coal mine dust and to protect the coal mine underground work personnel's physical and mental health, the present methods are "subtraction, drop, discharge, division" and individual protection, like coal seam water injection, wet operations, sprinkling water spray, spray foam and adding wetting agents, ventilation dust and dust collector, and others are followed (Wang, 2012; Jin, 2010; Li, 2005; Xiong, 2007; Zhao, 2005; Zhang, 2012). Deficiencies of these current measures are found at the scene: the spray is easily blocked; maintenance is difficult; spray atomization effect is not ideal; coal seam water injection operation process is relatively complex; foam dust removal cost is high; the dust catcher is relatively heavy and inconvenient to install and use. For the above reasons, the spray is used with self-exciting water-bath water-film dust remover cooperating with ventilation duct with Coanda effect (VDCE) under the long-pressure short-smoke ventilation mode in comprehensive dust control in 21007 comprehensive tunneling face of Chen Silou mine as research and experiment.

2 COMPREHENSIVE TUNNELING WORKING SURFACE PROFILES

The dip angle of the coal seam of comprehensive tunneling excavation face is 0–12°, an average of 7°, and the hydrostatic pressure is 2.37–3.50 MPa. Absolute gas emission of tunneling faces is 0.49~0.75 m³/min. Cross-section of roadway is trapezoidal and width×height = 4000 × 2600 mm. When coal seam is thin, it is appropriate to break end to ensure that roadway center height is not less than 2.6 m with floor construction level. When the coal seam thickness is more than 3.5 m, surplus coal can remain to ensure that roadway center height is not less than 2.6 m and not more than 3.5 m with floor construction level. The face used forced ventilation with a diameter of 0.6 mm and an air volume of 330 m³/min.

3 COMPREHENSIVE TUNNELING FACE DUST CONCENTRATION MEASUREMENT

3.1 Monitoring point arrangement

According to the dust concentration measurement procedures, combined with the actual, decorate sampling points in the return air side of roadway pavement breathing zone height from the machine driver, arrangement of measuring points is shown in Figure 1.

3.2 Comprehensive tunneling excavation face dust concentration measurement and analysis

Using a dust sampling instrument, AKFC - 92, some samples are collected and the sampling time is 2 min and the flow rate is 20 l/min. A dust concentration distribution curve is generated according to the test data, as shown in Figure 2.

From Figure 2, it can be seen that dust concentration is higher within 10 m distance of the driver in the face return air side. The total dust concentration in driver without measures is 1267.5 mg/m³. Respirable dust concentration is 1077.5 mg/m³, which is 126 times and 307 times more than the national standard, respectively; the dust concentration in the driver with the original spray is 875.5 mg/m³. Respirable dust concentration is 738.5 mg/m³, which is 87 times and 210 times more than the national standard, respectively; pressure air duct is hung over the right help, with the air flowing from the outlet at a high speed, which causes uneven local wind speed of roadway cross-section and some small vortex flow in front of the face. The dust blows into the return air side, leading to a high concentration of return air side. Dust concentration drops rapidly after driver. The dust particle size is larger, and gravity settling effect is apparent. Roadway wind speed is low, which is conducive to the settlement of dust.

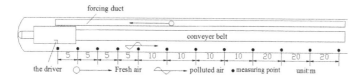

Figure 1.   Measurement points of dust concentration distribution at cutting coal.

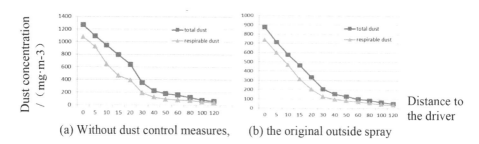

(a) Without dust control measures,   (b) the original outside spray

Figure 2.   Inertia effect diagram of single droplet to dust particle.

4 THE NUMERICAL SIMULATION RESEARCH ON DUST TRANSPORT

4.1 Geometric modeling and meshing

There are numerous equipment and people in small space of tunneling faces, which makes the dust area shape very complex and unable to make an accurate geometric model. So dust diffusion calculating seepage on the heading face range is properly simplified as follows: the cross-section of roadway is set to the shape of a trapezoid with the dimension of 4 m × 2.6 m; roadway length is100 m; dust collector export position is 22 m from face; roadway uses long-pressure short-smoke ventilation, with pressure air duct hanging on the right side of roadway. The diameter of pressure air duct is 0.6 m. The distance from the air duct axis ride height to the roadway floor is 1.8 m, and the distance from air duct outlet to the tunneling excavation face is7 m. The diameter of Aspirator suction duct is 0.5 m. The pumping air volume is 180 m³/min. In the roadway drivage GAMBIT software is used to establish geometric model and computing grid (Wang, 2007; Wang, 2006; Qin, 2011; Du, 2010; Hu, 2012; Hu, 2013), as shown in Figure 3.

4.2 The distribution regularity of flow field roadway

The wind flow field distribution and the direction of the wind generated by FLUENT is shown in Figures 4–6.

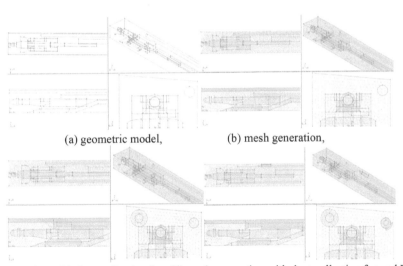

(a) geometric model,   (b) mesh generation,

(c) mesh generation with dust-collecting fan, (d) mesh generation with dust-collecting fan and VDCE

Figure 3.   Geometric model and mesh generation of excavation roadway.

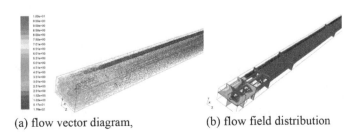

(a) flow vector diagram,   (b) flow field distribution

Figure 4.   Laneway airflow and flow field distribution at pressure-ventilating.

From Figure 4 it can be seen when air duct is jetted at a high speed from the right side of the tunneling roadway with forced ventilation, the most wind is oriented to the left of the roadway, hampered by face and wind speed to stabilize after the spread of a distance, and distributed evenly over the whole cross-section of roadway. Influenced by machine, the stability of high-speed wind is poorer within 20 m distance, and flow field distribution gradually stabilized after more than 20 m distance. Wind speed keeps at 0.4–0.5 m/s.

As illustrated in Figure 5, after joining dust catcher, the pressure air duct is absorbed and filtered, front-end wind exports a jet effect similar to that before under forced ventilation and exhaust inlet. The influence is reduced and a small suction scale wind field can form in front of face; part of the wind goes into the suction hood of dust-collecting fan; the others keep in the roadway.

From Figure 6, it can be concluded that, by adding the VDCE after the air duct, the most wind is discharged by side seam of VDCE and turns the high speed wind flow to the low speed of a rotating wind under the influence of the cross-section of roadway, which can form a contrarotating wind wall to prevent the dust source in the area of machine to rear diffusion of dust particles. At the same time, it guarantees that the wind flow field distribution is more homogeneous in the area behind machine, and the wind flow field in the right area is stable, and the wind flow field on the left side is improved. It formed a wide range of negative pressure area to suck the wind between machine and the face, which can avoid rotating vortices in the area and be conducive to the accumulation of dust in the area quickly and effectively.

4.3 Dust concentration distribution in driving roadway

The discrete phase model is turned on, based on the wind flow field. The dust source is joined separately in machine cutting head around and shovel plate position. Dust concentration distribution in the roadway computing diffusion condition is shown in Figures 7–9.

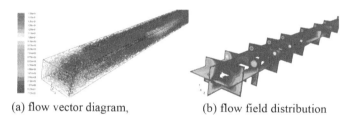

(a) flow vector diagram,   (b) flow field distribution

Figure 5.

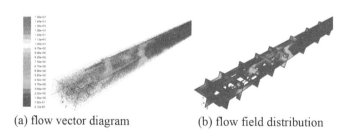

(a) flow vector diagram   (b) flow field distribution

Figure 6.   Laneway airflow and flow field distribution with dust collector and VDCE.

(a) Dust diffusion area, (b) breathing zone height, (c) different sections of the X direction, (d) pavement of return air side

Figure 7.   Dust concentration distribution in different section of roadway.

(a) Dust diffusion area, (b) breathing zone height, (c) different sections of X direction, (d) pavement of return air side

Figure 8.   Dust concentration distribution in different section of roadway with dust collector.

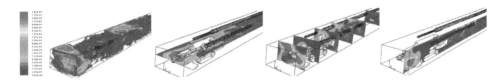

(a) Dust diffusion area, (b) breathing zone height, (c) different sections of X direction, (d) pavement of return air side

Figure 9.   Dust concentration distribution in different section of roadway with dust collector and VDCE.

From Figure 7, it can be seen that dust concentration around the cutting head is near the maximum in different sections of the roadway with the original dust concentration up to 1200 mg/m³; dust is in constant discharge and sedimentation in the wind flow along the whole section of roadway; jet zone lies between tunnel faces and the outlet duct, in which the dust mainly come from the wind itself contains and jet entrainment from eddy current area, and the dust concentration along the road is growing; the dust concentration is higher in the recirculation region and the eddy current region and falling along the road.

From Figure 8, it can be seen that after joining dust-collecting fan, the dust absorb aggregation and part of dust move to the downstream of the dust-collecting fan and then move back. Most of the dust no longer spread to the rear area of the roadway, but there are still some dust flow and spread to the rear in the return air side.

From Figure 9, it can be seen that after joining VDCE, most of the dust was well closed in the face area under the influence of rotating air curtain; at the same time, the suction hood is more effective in the steady air-flowing field. The high concentration dust is accumulated in the suction negative pressure zone, which improves the effect of the dust-collecting fan.

5   THE SCHEME AND APPLICATION OF DUST CONTROL

5.1   A self-excited bath water film dust catcher

A self-excited bath water film dust collector (SWDC) is a kind of scrubber using dust air directly of a certain kinetic energy impacting liquid in the SWDC to form liquid membrane and droplets, whose main structure is shown in Figure 10.

The dust air enters the SWDC through the air inlet and turns down to impact water so that large-diameter dust grain due to inertia effect falls into the water tank, and smaller dust particles with air at high speed are captured and settled by a large number of water droplets prompted when it goes through the bending passage between the guide vane. The dust air with water flow can form water films on the inner wall of SWDC and the upper and lower guide vane in the centrifugal force to trap dust and sedimentation. Then, the water droplets should be separated from the air by dehydrator of SWDC and discharged through an axial flow fan. The mensuration of breathing dust removal efficiency is above 96%.

5.2 Dust control mechanism of ventilation duct with Coanda effect

Ventilation duct with Coanda effect (VDCE) can make the original axial wind pressure in air duct supply excavation face change along the rotation of the wind tunnel walls with a certain rotational speed blowing the tunnel wall and the whole cross-section of roadway, and advancing face. The wind forms a spiral linear flow under the dust suction wind flow and the air curtain formed in front of machine driver to make dust air through SWDC to purify instead of spreading outward, which improves the efficiency of SWDC, as shown in Figure 11.

5.3 Field application and effect

According to field conditions and the procedures requirement, the excavation face adopts long pressure short smoke ventilation. The VDCE is installed at the outlet of forcing duct. The outlet duct is 8 m apart from the face. The suction duct is 0.5 m in diameter. Suction hood is arranged by boring and cutting arm rear of the machine. The SWDC is borne by car wheel on the slide rail connecting to the second shipment (Ding,2015;Li,2015), as shown in Figure 12.

At the same time, in order to improve the insufficiency of traditional spray, the structure of nozzle, installation location, number, arrangement and the fog flow direction, and others are improved. A new type air–water spray can form uniform stability of bubble two-phase flow in front of the nozzle (Yang, 2005; Wu, 2003), which has a good atomization effect, appropriate particle diameter size, large jet momentum and long distance, wide coverage, and small water pressure requirements. It is suitable for complicated conditions of coal mine.

The dust concentration and concentration distribution curve, drawing different dust suppression in coal face, is shown in Figure 13.

After the improvement of spray, total dust concentration is 383.5 mg/m³, and respirable dust concentration is 286.5 mg/m³ in the driver where dust rate is more than 60%. A considerable amount of dust is reduced in coal-cutting operation, generated in the tunneling excavation face.

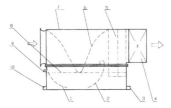

Figure 10. Dust collector structure diagram of air duct.

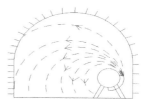

Figure 11. Out wind state diagram of air duct of VDCE.

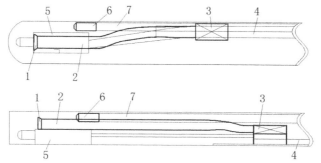

1 – exhaust inlet; 2 – exhaust duct; 3 – SWDC; 4 – conveyer belt;
5 – heading machine; 6 – VDCE; 7 – forcing duct

Figure 12. Dust removal equipment layout.

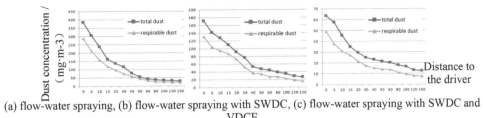

(a) flow-water spraying, (b) flow-water spraying with SWDC, (c) flow-water spraying with SWDC and VDCE

Figure 13.   Dust concentration distribution along the way.

Using spray with SWDC, total dust concentration is 171.5 mg/m³, and respirable dust concentration is 131.5 mg/m³ in the driver where dust rate is more than 80%. Use of SWDC is more effective for dust controlling than using spray only.

Reduction of the dust effect is remarkable using comprehensive technology of dust control. Total dust concentration is 63.5 mg/m³, and respirable dust concentration is 49 mg/m³ in the driver where total dust rate is as high as 93.4%. The respirable dust rate is as high as 93.7%.

6   CONCLUSION

1. There is a high dust concentration in comprehensive tunneling face coal cutting process, thus a single method cannot achieve the ideal dedusting effect. The traditional outside spray dust effect is 30–40%, which cannot achieve the requirement of the occupational disease prevention and control of coal mine workers.
2. The flow field and dust distribution of comprehensive tunneling excavation face is studied adopting the method of three-dimensional numerical simulation. The results show that the high-speed wind discharging from the outlet of forcing duct would create a great obstruction to the flow field of comprehensive tunneling excavation face; the large amount of dust is produced in cutting process flow and deposits backward with the wind migration of face; the air volume of SWDC should not be too big to prevent circulation wind; at the same time the suction hood range is limited under the compression wind disturbance, so the formation of negative pressure zone range is smaller; VDCE can effectively improve the flow field of face and spin forward as the wind wall from the driver, which will form a wide range of negative pressure area together with SWDC where the high concentrations dust will be inhaled and treated effectively.
3. The application of the field shows that the dust removal effect is significant with the total dust and respirable dust suppression efficiency over 87% and greatly improves the underground work environment, providing guidance and reference to other coal mine tunneling face dust control.

REFERENCES

DING,H.C. et al.2015.Research on Dust Migration Law and Its Removal Technology in Fully Mechanized Working Face. Safety and Environmental Engineering 22(4):82–87.
DU,C.F. et al.2010.Numerical simulations of dust distribution in a fully mechanized excavation face with far-pressing-near-absorption ventilation. Journal of University of Science and Technology Beijing 32(8):957–961.
HU,F.K. et al.2012.Analysis on dust migration on heading face based on CFD numerical simulation. China Coal 38(6):94–98.
HU,F.K. et al.2013.Unsteady simulation analysis of dust movement law at driving face based on CFD. China Coal 39(3):105–108.
JIN,L.Z. et al. 2010. The mine dust control theory. Beijing: Science Press.
LI,H.W. 2005.Comprehensive control of respirable dust in coal mines. China Safety Science Journal 15(7):67–69.

LI,S.F.2015. Dust Control Technology at Fully Mechanized Working Face Based on Wet Collector and Wall-attached Fan Drum. Safety in Coal Mines 46(6):57–59.

QIN,Y.P. 2011. et al. Numerical simulation of dust migration and study on dust removal modes with the forced ventilation shunt in a fully mechanized workface. Journal of University of Science and Technology Beijing 33(7):790–794.

WANG,D.M. 2012. mine ventilation and safety.Xuzhou: China University of Mining and Technology Press.

WANG,X.Z. et al. 2006.Numerical simulation of distribution regularities of dust concentration during the ventilation process of coal drift driving with exhaust ventilation. Journal of Safety Science and Technology 2(5):24–28.

WANG,X.Z. et al. 2007.Numerical simulation of distribution regularities of dust concentration during the ventilation process of coal roadway driving.Journal of China Coal Society 32(4):386–390.

WU,D.H.2003.Analysis and evaluation on the influence characteristics of jet nozzle. Tianjin: University Of Science and Technology Of Tianjin.

XIONG, A.J. 2007.Discussion on comprehensive dustproof in mine.Hebei coal 1:25–27.

YANG,X.J.&YAN,H.r.2005.Experimental study on fan nozzle. Chinese Agricultural Mechanization 1:39–42.

ZHANG,Y.K. 2012.Study on dust control technology with foam in fully mechanized workface. China Safety Science Journal 22(2):151–156.

ZHAO,C.G. 2005.Techonlogy and effectiveness analysis of coal seam water flooding. Coal Science & Techonology magazine 1:45–47.

Hydraulic Engineering IV – Xie (Ed.)
© 2016 Taylor & Francis Group, London, ISBN 978-1-138-02948-4

Hydrothermal synthesis of gypsum whisker by industrial phosphogypsum washed with water

Chuanyang Zheng, Xiaomin Jiang, Yangqun Zhou, Tianyao Qi & Tingting Hou
Department of Biology and Environment Engineering, Hefei University, Hefei, Anhui, China

Daming Gao
Department of Chemistry and Materials Engineering, Hefei University, Hefei, Anhui, China

ABSTRACT: Gypsum whisker was hydrothermally synthesized by purification of phosphogypsum, washed with water and screened with vibration, which removed phosphoric acid compound, halide ion and water-solvent compounds and abandoned the weak radioactive compounds with uranium and thorium. The resultant product has a large length-diameter ratio with a high tensile strength, modulus of elasticity and flame retardant and good morphology with the same radium in a bunch, which is a very broad prospect in resin, rubber, paint, paper, friction materials, building materials, sealing materials and insulation materials.

1 INTRODUCTION

Phosphogypsum (PG) is a solid waste discharged from the production of phosphoric acid by the wet acid method, which is composed primarily of calcium sulfate dehydrate and partially some impurities, such as phosphate, fluorides, sulfate ions, organic compounds and weak radioactive compounds with uranium and thorium (Singh, 2002; Du et al., 2010; Azabou et al., 2005; Haridasan et al., 2001; Altun, 2004). An amount of 1-t phosphoric acid prepared in chemical process results in a discharge of about 4.5–5 t of PG. Each year 280 million tons of phosphorus gypsum is produced at international level, of which China's annual waste of PG is 50 million tons, and it is growing at a rate of about 15% a year; however, according to the statistics, the PG utilization rate is only about 20%, and utilization rate lags further than that of PG produced. At present, there are many researches on PG resource utilization, which is mainly used for making sulfuric acid, production of cement and cementing materials, and as soil conditioner and feed and others. But a large number of PG stacking is a serious phenomenon at present. This will not only occupy land, but also can cause the pollution of soil and water. Utilization of PG encountered the problem of bottleneck for pretreatment by purification and transfer to gypsum whisker (Deng et al., 2004; Ghosh, 2010; Cichy et al., 2013; Gong et al., 2010; Wang et al., 2006; Shen et al., 2012). Herein, PG was washed with water and screened with vibration to remove the impurities and to provide more possibilities for the comprehensive utilization of PG. The PG and synthesized gypsum whisker were characterized using X-Ray Fluorescence (XRF), X-Ray Diffraction (XRD) and Scanning Electron Microscope (SEM) to test the physical properties, and the main analysis of characteristics of the chemical composition and morphology displayed an information of PG and gypsum on the morphology, composition, to further provide the technology support for the comprehensive utilization of PG and gypsum.

2 EXPERIMENTAL DETAILS

2.1 *Materials*

PG samples [main content $(CaSO_4 \cdot 2H_2O) \geq 85$ mass%] were provided by Anhui Zhongyuan Chemical Science and Technology Co., Ltd. Deionized water (18.2 MΩ cm) was self-made from our lab. Industrial-grade lime powder $(CaO \geq 95$ mass%) and ethanol were purchased from Sinopharm Chemical Reagent Co., Ltd. The purities of the chemicals were provided as per the manufacture's specifications.

2.2 *Characterizations*

The morphology of PG was examined by a SU8010 field-emission scanning electron microscope. XRD patterns were recorded using an X-ray diffraction DT-3500 analyzer (Dandong, China).

2.3 *Experimental process*

An amount of 5-g solid waste PG was weighed and added into the vessel (liquid-to-solid ratio of 6:1) with a mixture, having a concentration of 0.4% of lime water. The mixtures were stirred at 250 rpm for 6 min and statically precipitated for 1 h at room temperature. The floating matters on the surface of upper clear liquid were removed, and the slurry was mixed properly. The uniform slurry was moved into a filtration film for dehydration treatment by a circulation-type vacuum pump in a tank. After dehydration, the PG was placed in an air blast drying oven, in which the temperature was kept constant by electric heating for 12 h at 75°C to obtain the purified PG.

   The mass fraction of 2–3% of a PG solution was introduced into a stainless steel reaction vessel with an inner Teflon lining pot. After sealing the reactor, the reactant was used for 10 min at room temperature. The reactor was put into the oven for crystallization reaction for 3 h at 145°C, and the resultant product was washed with deionized water, centrifuged and dried for 12 h at 45°C in a vacuum oven to obtain gypsum whiskers.

3 RESULTS AND DISCUSSION

3.1 *XRF analysis of raw material phosphogypsum*

Untreated PG samples were measured by an XRF spectrometer. An appropriate amount of PG samples were placed into a high-temperature tube furnace to calcine at a certain temperature. Alkaline absorption liquid may absorb volatile substance using nitrogen as the carrier gas. For the determination of residue after calcination and absorption of water-soluble phosphorus and fluorine, phosphorus vanadium molybdenum yellow double wavelengths spectrophotometers approach was to detect water-soluble phosphorus content, and selective electrode method was used to detect water-soluble fluorine content (JC/T 2073–2011). The main chemical composition and content of phosphorus gypsum is shown in Table 1. The determination method uses reference standards 5484–2000GB/T, 23456–2009GC/T and 2073–2011GB.

Table 1. Bulk chemical components of phosphogypsum.

Components	SiO_2	Fe_2O_3	CaO	MgO	SO_3	Al_2O_3	TiO_2	$P_2O_5{}^*$	F⁻★	Ignition loss
m%	4.28	0.13	32.16	0.29	44.37	0.35	0.03	2.62	0.23	15.54

* Soluble P_2O_5: 0.575%; ★ Soluble F⁻: 0.134%.

Table 1 displayed raw PG materials mainly consisting of CaO and SO_3, followed by SiO_2 and P_2O_5, in addition to a small amount of TiO_2, MgO and other materials; thus, the PG content is more complex, including many types of harmful impurities.

3.2 XRD analysis

Figure 1 represents the XRD patterns of before (a) and after (b) purification of PG. According to the analysis of the Jade Software, Figure 1a shows that unpurified PG mainly contains $CaSO_4 \cdot 2H_2O$ (PDF 33–0311), and the diffraction peak intensity is high. There are different sizes of peaks of diffraction at the bottom of the figure. These features are more obvious for SiO_2 (PDF 44–069) and P_2O_5 (PDF 23–1301) peaks. Through the comparison of the standard card of PG XRD diagram with the powder diffraction database, it can be learned that XRD peaks of PG are mainly composed of characteristic peak of $CaSO_4 \cdot 2H_2O$ (according to the PDF06–0047standard card) and a small amount of SiO_2 characteristic peaks (according to the PDF03–0420standard card). Figure 1b displays that purified PG mainly contains $CaSO_4 \cdot 2H_2O$ (PDF 33–0311), the different sizes of peaks of diffraction at the bottom of the figure are less and small relative to Figure 1a, and there are almost no diffraction peaks for P_2O_5; the SEM image of the PG surface further shows that the PG surface is attached to small and irregular soluble phosphorus, fluorine, organic matter and other impurity particles. To some extent, washing pretreatment removed harmful impurities, soluble phosphorus and water-soluble fluorine from PG, when pH value of waste liquid was modulated by adding high-concentrated lime water, insoluble precipitates of Ca_3HPO_4 and CaF_2 were formed, which reduced the influence of impurities on the hydrothermal synthesis whisker (Shen et al., 2012).

3.3 Morphology analysis before and after PG purified

Figure 2a shows that SEM morphology of the unpurified PG is a parallel quadrilateral, also with a lot of debris, dominated by the rules of the parallel quadrilateral plate. Figure 2b shows a single phosphorus gypsum particle magnified by 6000 times, which is a rule of the parallel quadrilateral plate; its surface is compact and contains a large number of impurity particles. The light gray morphology of PG mainly exists in the rules of parallel quadrilateral plate shape, where most of the crystalline form is regular and uniform, and the plate surface is attached to small and irregular soluble phosphorus, fluorine and organic matter, such as impurity particles.

Figures 2c and d shows the SEM image of the purified PG at1000 and 6000 times magnification, respectively. In comparison with Figure 2a and b, the surface of parallelogram-shaped plate is smooth and clear, and there is no obvious attachment material. This shows

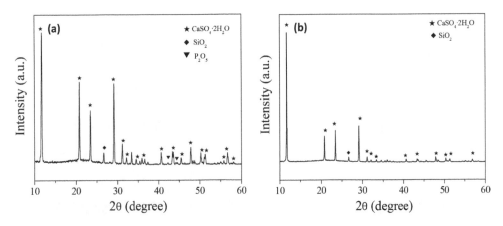

Figure 1. XRD patterns of before (a) and after (b) purification of phosphogypsum purified.

that water washing pretreatment, to a certain extent, removed $CaHPO_4$, CaF_2 and organic impurities present in the raw materials of PG.

3.4 *Morphology analysis of gypsum whiskers that is hydrothermally synthesized from unpurified and purified PG*

Figure 3a and b shows the SEM image of morphology of gypsum whiskers hydrothermally synthesized from unpurified and purified PG, respectively. As shown in Figure 3, the gypsum whisker from purified PG is much better than that of unpurified PG. The gypsum whisker from purified PG displayed a high long diameter ratio and is relatively regular. The purified PG synthesized whisker with glycerol and water mixture is of a long diameter ratio and is in a good shape, which shows that a certain amount of a glycerol structure directing agent is required to promote the one-way hydrothermal growth of gypsum whisker.

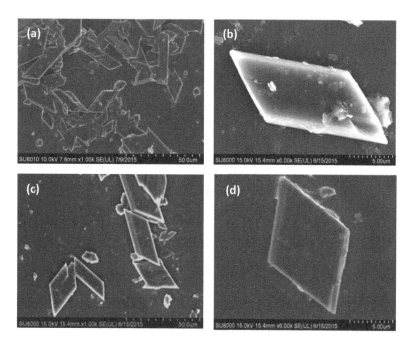

Figure 2. SEM images of before (a and b) and after (c and d) purification of phosphogypsum.

Figure 3. SEM images of before (a) and after (b) phosphogypsum-purified hydrothermally synthesized whisker, respectively.

242

4 CONCLUSIONS

The raw material of PG can be found in the rules of the parallelogram plate, and a small amount of finely, and the main ingredient of PG is $CaSO_4 \cdot H_2O$, meanwhile it also contains phosphorus, fluorine, silicon and other impurities. The surface of purified PG is relatively smooth and obviously don't have attachments because the process of water purification which removed the harmful impurities present in the raw material of phosphorus gypsum to a certain extent. The morphology of the purified PG synthesized gypsum whisker with glycerol and water mixture is better than that of unpurified PG. The main reason is that a minute amount of structure oriented agent propylene glycol can promote one-dimensional orientation growth of the calcium sulfate whisker.

REFERENCES

Altun, I.A. et al. 2004. Utilization of weathered phosphogypsum as set retarder in Portland cement. *Cement and Concrete Research*34(4): 677–680.

Azabou, S etal.2005. Sulfate reduction from phosphogypsum using a mixed culture of sulfate-reducing bacteria. *International Biodeterioration & Biodegradation* 56(4): 236–242.

Cichy, B. et al. 2013. Phosphogypsum management. world and polish practice. *Przemysl Chemiczny* 92(7): 1336–1340.

Deng, Z.T. et al.2004.Effect of phosphorus gypsum cake on the reduction of phosphorus in water washing and Re understanding of the practice. *Phosphate and Compound Fertilizer*19(3): 24–26.

Du L.S. et al.2010. Utilization and recovery of phosphorus gypsum. *Chemical Technology and Development* 39(4): 25–28.

Ghosh, A. 2010. Durability of lime-fly ash stabilized soil activated by calcined phosphogypsum. *Journal of Materials in Civil Engineering* 22(4): 343–351.

Gong, H.Y. et al. 2010. Progress in the study of the effect of impurities on the crystallization process of solution. *Chemical and biological engineering*27(3): 9–12.

Haridasan P. P et al. 2001 Natural radio nuclides in the aquatic environment of a phosphogypsum disposal area. *Journal of Environmental Radioactivity* 53(2): 155–165.

Shen, W.G. et al. 2012. Utilization of solidified phosphogypsum as Portland cement retarder. *Journal of Material Cycles and Waste Management* 14(3): 228–233.

Singh, M. 2002. Treating waste phosphogypsum for cement and plaster manufacture. *Cement and Concrete Research* 32(7): 1033–1038.

Wang, L. et al.2006. Study on Preparation of calcium sulfate whisker by hydrothermal method and its crystal morphology. *Materials science and technology* 14(6): 626–629.

Hydraulic Engineering IV – Xie (Ed.)
© *2016 Taylor & Francis Group, London, ISBN 978-1-138-02948-4*

Effects of driving lane width on three-arm intersections on drivers' speed in urban areas

R. Ziolkowski
Bialystok University of Technology, Bialystok, Poland

ABSTRACT: Driving speed is one of the most important factors influencing road safety. Speed not only affects the severity of a crash but is also related to the risk of being involved in a crash. Although human factor is considered to be the most important one, road geometry is getting more concerned in safety analyses. In Poland, every year around 70% of all traffic accidents occur in urban areas, from which most is recorded at intersections as a consequence of excessive speed. This paper investigates the influence of geometry characteristics in terms of driving lane width, channelization and traffic organization on average speed under free-flow conditions. The speed measurements were undertaken in Bialystok, Poland on approaching sections of three-arm priority intersections situated in an urban area. Speed data were collected under free-flow conditions with the use of radar speed gun. To establish the statistical efficiency, the variance analyses were applied.

1 INTRODUCTION

The road geometry design considerations include a number of factors related to functional and technical road classification, length of straight and curved sections, expected traffic volume, cross section with a number of driving lanes and widths, junction density, vehicle characteristic, environmental considerations, design speed and drivers' expectations. Among these all factors in Poland, speed still remains a major one in terms of traffic conditions and safety. The importance of speed is undisputed in both rural and urban areas and is reflected in numerous publications in relation to drivers' behaviour and road characteristics. Excessive speed has been already pointed to be a primary problem in traffic by Fildes and Lee (1993), and the problem of exceeding drivers in rural and urban areas has been investigated in many countries (Aljanahi et al., 2001; Canel and Nouvier, 2005; Cottrell et al., 2005).

Roadway geometric effects were explored in scope of their safety influence and road accident frequency and severity. Milton and Mannering (1998) analysed and compared the annual accident frequency on short and long sections of principal city arterials and found that short sections are less dangerous than long ones. Road curvature and its association with traffic crashes was also studied by Haynes et al. (2007). Their investigations developed a number of measures for road curvature and its protective influence on road safety. Also, the study conducted by Wang et al. (2009) in scope of road curvature confirmed an inverse relationship between road curvature and road accidents. However, findings of Abdel-Aty and Radwan (2000) contradict those of Haynes. According to their research results, the degree of curve increases the number of accidents. These discrepant findings may only confirm that road geometric effects both speed and safety.

Safety analyses comprise many speed aspects and alternative solutions to be implemented in order to effectively manage and control vehicle speed. Amongst main aspects of road infrastructure that influence speed choice in urban areas are lane width, road width, road shoulder width, number of lanes, signage, channelization, traffic flow and roadside environment (Mannering, 2009). Elements related to cross section, such as lane width and number of lanes, have the strongest influence on driver's perceptions of safety and travel speeds

(Rosey et al., 2009). The research results of lane width on driving speed are not consistent. According to Urban Street Design Guide (2013), narrower streets and travel lanes not only are correlated with lower speeds but also provide reduced crossing distances and shorter signal cycles. Anuj et al. (2015) also stated that lane width design should also consider a specific location of a street segment, existing speed limit and traffic characteristic.

The main objective is to investigate the speed and its dependency on chosen geometry characteristics of three-arm priority intersections located in urban areas. Detailed analyses have focused on lane width, channelization islands and turn lanes presence on drivers' speed on intersection approach sections under free-flow traffic conditions.

2 METHODOLOGY

2.1 Site characteristic

The study was conducted on approach sections of three-arm priority intersections located in Bialystok, Poland. The speed measurements were taken in two sections located at a distance of 150 m and 40 m from the intersection.

Three-arm priority intersections, varied in terms of the presence of turn lanes and channelization islands, were chosen for detailed investigations. Lane width varied from 2.75 to 4.5 m depending on road width and traffic organization on a specific intersection. Detailed characteristic including intersection facilities is presented in Table 1. Each intersection was labelled with a digit and a letter symbol. Intersections 1, 2, 3 and 5 are located in built-up area, whereas intersections 6 and 7 are situated in city outskirts. Existing speed limit in urban areas is 50 km/h with the exception given to the intersection 5 where local speed limit is lowered to 40 km/h. The localization of measurement points was chosen to achieve average speed data in two characteristic sections. The first section was located at a distance of 150 m from the investigated intersection to provide speed data characteristic for mid-block segments. The second point was placed about 40 m from the yield line at a short distance from a pedestrian crossing to reflect possible influence of pedestrian facility presence on drivers' speed.

2.2 Speed measurement

The spot speed of passing vehicles was measured by a radar gun. The operator with the gun was always hidden from traffic in order to minimize the effects of his presence on drivers and, consequently, on travel speed. Only the speeds of free-flowing passenger cars were measured. To minimize the effect of a surveyor with the radar gun on drivers' speed choice, the surveyor sat in a car situated by the roadside.

Table 1. Basic intersection characteristics.

Intersection feature	Kawaleryjska/ Sloneczna (1)		Wroclawska/ Niepodleglosci (2)			Ciolkowskiego/ Plazowa (3)			Dolistowska/ Wloscianska (5)		Mazowiecka/ Brukowa (6)		Mazowiecka/ Horodniany (7)	
Inlet	1A	1B	2A	2B		3A	3B		5A	5B	6A	6B	7A	7B
Turning movement	L	R	R	L	R	R	L/R	R	R	L	L	R	L	R
Channelization	Yes	No	Yes		No	No			No		No		Yes	No
Lane width (150 m)	4.5	4.5	3.3	3.75	3.75	4.5	3.5	3.0	3.5	3.5	3.5	3.5	3.5	3.5
Lane width (50 m)	2.75	2.75	4.5	3.5	3.5	4.5	3.5	3.0	3.5	3.5	3.5	3.5	3.5	3.5

3 RESULTS AND DISCUSSION

All data were collected in daylight, on a dry road. For each turning movement, a number of 60–80 vehicles were collected.

The summary characteristics of the key speed parameters are presented in Table 2.

The speed data given in Table 2 reveal that there exists a clear speed diversity depending on such factors such as

- the distance from the yield line (Fig. 1a),
- the existence of turn lanes (Fig. 1b),
- planned turning movement (Fig. 1c) and
- traffic organization (presence of turn lanes, median islands, priority of the inlet, lane width – Fig. 1d).

Conducted variance analyses have confirmed statistically essential differences between average speeds presented in Figure 1. In the case of a lane width influence, such a difference was stated between values recorded on roads with lane width 2.75 and roads with a wider lane width.

Generally, in line with the expectations, the higher the distance from an intersection, the higher speeds drivers travelled. At a distance of 150 m from tested intersections, average speeds do not differ statistically between themselves at a significant level, $p = 0.05$. However, those results supported by 85th percentile values confirm a serious problem, previously stated by the author Ziolkowski (2012), with speeding drivers. It is especially distressing when 85th percentile is considered at a short distance from the intersection, shortly before pedestrian crossings. In Poland, the existing speed limit in urban areas is 50 km/h, and the values recorded on intersection approaching sections at a distance of 40 m have many times exceeded the limit reaching values from 39.9 km/h to 60.2 km/h, regardless of the lane widths. On the other hand, when analyses are focused on the influence of turn lanes presence on average speed, it occurs that they have a positive impact on lowering vehicles' average speed (Fig. 1b). Such an effect was proved for left-turn movements along primary approach sections. Average speed values recorded on section with a turn lane facility (Mazowiecka/ Horodniany_7 AL) were lower by 9.4% than average values recorded on the section without such a facility (Mazowiecka/Brukowa_6 AL). A similar relationship can be observed if one considers average speeds' dependence on the approach section priority type (Fig. 1c). Drivers approaching intersection along a priority road drive faster by about 5.5 km/h (11.8%) than those travelling along a secondary road. Data presented in Figure 1d show that lane width

Table 2. Key speed parameters for the investigated intersections.

	Kawaleryjska/ Sloneczna		Wroclawska/ Niepodleglosci		Ciolkowskiego/ Plazowa			Dolistowska/ Wloscianska**		Mazowiecka/ Brukowa			Mazowiecka/ Horodniany	
Inlet	1A	1B	2A	2B	3A	3B		5A	5B	6A	6B		7A	7B
Turning movement	L	R	R	L	R	R	L/R	R	R	L	L	R	L	R
Average speed ($V_{avg,150}$)	52.2	51.3	49.1	53.8	52.9	44.5	41.2	52.2	46	49.8	54.2	66.4	47.2	51
Average speed ($V_{avg,50}$)	42.7	41.1	44.7	48.7	48.7	40.8	43.3	48.8	39.9	40.9	46.7	60.2	42.3	44.3
$V_{85,150}$	59	63.8	56	65	60.1	52	49.5	58.4	53.8	58	59.4	73.6	51	56
$V_{85,50}$	48	48.3	50	55	58	48	50.5	54	47	48	51.2	67	46	48

*The average value increased.
**Existing speed limit was 40 km/h.

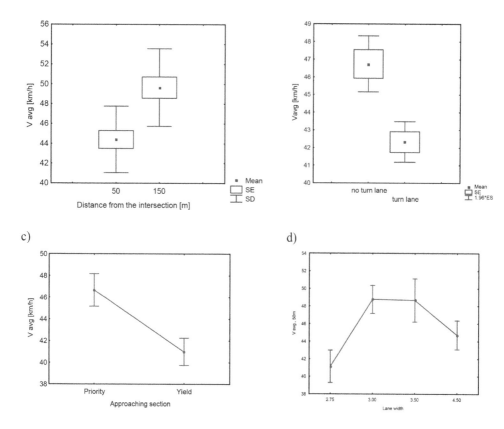

Figure 1. Average speeds in relation to (a) the distance from the intersection, (b) presence of a turn lane on approaching section, (c) priority at the intersection and (d) lane width.

also has significant influences on the average speed. The highest average speed values at a distance of 40 m from the intersections 1–5 were highest on streets with lane widths of 3.0 and 3.5 m (48.3 km/h), whereas lower average speeds were observed on streets with a traffic lane width of 4.5 m (44.4 km/h), and the lowest values were registered on streets with driving lane width of 2.75 m (41.3 km/h).

From the data shown in Table 1, it appears that the angle of intersecting road axes significantly lower from 90° (intersection 1), although not suggested due to adverse visibility, forces drivers for more careful and slower driving. Average approaching speeds at that intersection were lower than the values recorded in vicinity of the other intersections, with the exception given to intersection Dolistowska/Wloscianska where enforced speed limit is 40 km/h. In addition, conducted analysis of the influence of turn movement on average speed did not show any statistical significance on intersection approaches with such facilities.

3.1 Conclusions

The general goal of this study was to evaluate average speeds in relation to different driving conditions associated with geometry features of three-arm intersections. In Poland, most accidents occur in vicinity of intersections, and in this regard a remarkable observation was the excessive speed, which was registered not only on mid-block sections (150 m away from the intersection), but especially at a short distance (40 m) from the intersection. Achieved results allow us to draw the conclusion that designing of turn lanes can entail a positive effect on lowering approaching speed.

Keeping in mind the influence of speed on consequences of accidents, a particular consideration must be given to effective speed management at short distances from the intersection's entry line. Because of the presence of pedestrian crossings and vulnerable users and potential conflicts, those areas require special consideration during a design procedure. From the investigations, it results that intersections in city outskirts may require additional attention due to notably high approaching speeds. It may be reasonable in terms of safety considerations to design raised median islands at intersection, as their presence (Mazowiecka/Horodniany) may significantly lower drivers' speeds. It may also be interesting to deepen future research towards drivers' dynamic behaviour in vicinity of intersections.

REFERENCES

Anuj S., Wei Li M.S., Mo Zhao M.S., Laurence R.R. (2015)Safety and Operational Analysis of Lane Widths in Mid-Block Segments and Intersection Approaches in the Urban Environment in Nebraska. Nebraska Transportation Center.

Abdel-Aty M.A., Radwan A.E. 2000. Modelling traffic accident occurrence and involvement. Accident Analysis and Prevention 32 (5), 633–642.

Aljanahi A.A.M.; Rhodes A.H.; Metcalfe A.V. 2001. Speed, speed limits and road traffic accidents under free flow conditions, Accident Analyses and prevention (33): 585–597.

Canel A., Nouvier J. (2005) Road safety and automatic enforcement in France:results and outlook. Routes/Roads (1), 54–61.

Fildes B., Lee S. (1993). The Speed Review: Road Environment, Behaviour, Speed Limits, Enforcement and Crashes. Federal Office of Road Safety, Report CR127, Department of Transport and Communications, Canberra.

Haynes R., Jones A., Kennedy V., Harvey I., Jewell T. 2007. District variations in road curvature in England and Wales and their association with road-traffic crashes. Environment and Planning A 39 (2). 300–307.

Milton J., Mannering F. 1998. The relationship among highway geometrics, traffic-related elements and motor-vehicle accident frequencies. Transportation 25 (4), 395–413.

Mannering F. (2009). An empirical analysis of drivers perceptions of the relationships between speed limits and safety. Transportation research Part F, Traffic psychology and behaviour 12 (2), 99–106.

Rosey F., Auberlet J., Moisan O., Dupre G. (2009). Impact of narrower lane width comparison between fixed-based simulator and real data. Journal of the Transportation Research Board No 2138, Washington, DC: Transportation Research Boardof the national Academies; 112–119.

Urban Street Design Guide, National Association of City Transportation Officials, 2013.

Wang C., Quddus M., Ison S. 2009a. The effects of area-wide road speed and curvature on traffic casualties in England. Journal of Transport Geography 17 (5), 385–395.

Ziolkowski, R.2012. Wpływ środków uspokojenia ruchu na prędkość pojazdów w warunkach miejskich [Influence of traffic calming measures on speed in cities], Przeglad Komunikacyjny (4): 24–30.

Hydraulic Engineering IV – Xie (Ed.)
© 2016 Taylor & Francis Group, London, ISBN 978-1-138-02948-4

Simplified simulation of flows with turbulent macrostructure

O.M. Gumen
National Technical University of Ukraine "Kyiv Polytechnic Institute", Kiev, Ukraine

V.B. Dovgaliuk & V.O. Mileikovskyi
Kyiv National University of Construction and Architecture, Kiev, Ukraine

ABSTRACT: We propose an approach for simplified simulation of turbulent subsonic flows with macrostructure such as jets and boundary layers between flows. Usually, we use Large Eddy Simulation (LES) by the same process as physical experiments but with virtual 3D models. Professor A. Tkachuk (Kyiv National University of Construction and Architecture) developed a theory for simulation of flows with small-scale turbulent structure. Based on the small influence of viscosity, he said that such flow can be considered as ideal liquid flow with foreign bodies—vortexes. We extend the theory for large-scale vorticity. This vorticity is more ordered because of geometric compatibility. We can simulate it using geometric and kinematic analysis avoiding differential equations. This work has a final list of assumptions and as an example results for jets in flows. The results can be used in atmospheric or reservoir flow simulation and calculation of energy-efficient and environmentally safe HVAC and sanitary technical systems.

1 INTRODUCTION

Most flows in atmosphere, water reservoirs, and also in HVAC or sanitary technical systems are turbulent. The flows in pipes, ducts, or in immediate proximity to solid surfaces have only small-scale vorticity. In submerged flows or between flows, there is large-scale vorticity. For simulation of such flows, there are (Li, Gong, Wang, Zhang, Wei & Shu 2016; Cheviron & Moussa 2015; Arifin, Pasek & Eddy 2015; Stepanov, Pešenjanski & Spasojević 2015; Lee 2015; Kiczko, Kubrak, & Kubrak 2015; Bose 2015; Masters, Williams, Croft, Togneri, Edmunds, Zangiabadi, Fairley & Karunarathna 2015; Tian, Song, VanZwieten & Pyakurel 2015; Toja-Silva, Peralta, Lopez-Garcia, Navarro & Cruz 2015; Ismail & Batalha 2015; Kwang-Jun, Seunghyun, Jaekwon, Taegu, Yeong-Yeon, Haeseong & Suak-Ho 2015; Sungwook & Booki 2015; Alaimo, Esposito, Messineo, Orlando & Tumino 2015; Jianxi 2015; Anggiansyah & Prabowo 2014; Tongpun, Bumrungthaichaichan & Wattananusorn 2014; Bahirat & Joshi 2014; Jing, Li, Guo, Zhu & Li 2014; Ali, Megri, Dellenback & Yu 2014) different mathematical models (k–ε, RANS, etc.) and CFD software (free and commercial). In 1963, the American meteorologist and the first director of Geophysical Fluid Dynamic Laboratory of National Oceanic and Atmospheric Administration, Joseph Smagorinsky (1924–2005), proposed Large Eddy Simulation (LES) model (Smagorinsky 1963). It is based on the Navier-Stokes equation with low-pass filtering. It provides good results (TienPhuc, ZhengQi & Zhen 2016; Warzecha, Merder & Warzecha 2015; Nakayama, Takemi & Nagai 1915) because large eddies have incomparably more energy than small-scale vorticity and the last one may be filtered out (by low-pass filtering). All simulation approaches described below require the same procedure as physical experiments but using a virtual 3D model. Also for the best quality results, we need more computational resources and time.

Andrey Tkachuk (1928–2002), the professor of Heat Gas Supply and Ventilation Department of Kyiv National University of Construction and Architecture, developed a new approach for wall boundary layers (Tkachuk 2001). In a viscous layer that contacts with a wall, there is counteraction of viscous forces and forward movement. This counteraction forms initial vortex cords that move to turbulent core of the wall boundary layer by Magnus forces. In the core, these vortexes cause secondary vorticity without a direct influence of viscosity. So, the flow in the viscous layer can be assumed as a vortex cord film and the flow in the turbulent core—as ideal liquid flow. First flow acts as "singularities" that produce second one. The last flow can be calculated by ideal flow laws. As a result, Tkachuk theoretically proved not only logarithmic but also the experimental power velocity profile. He also obtained Darcy coefficient.

The similar approach can be used for large-scale vorticity.

2 ASSUMPTIONS FOR LARGE-SCALE VORTICITY

Subsonic Jets and boundary layers between subsonic flows have large-scale vortexes. This fact is known from ancient time (from the first human-obtained fire). The large-scale vortexes (puffs) stain by smoke particles and stay visible. Analysis of visual researches (including that of the author), geometry, and kinematic characteristics of the macrostructure lets us assume the following:

– a jet at the enough distance from inlet opening (slot) can be replaced by the equivalent Tolmien source from infinitely small opening (slot);
– the puffs may be considered as round (i.e., cylindrical vortex cord);
– distance between the puffs is small with respect to the puffs size, so the puffs may be enlarged up to touch;
– the puffs roll by free boundary(ies) of a boundary layer;
– in a free jet the line (surface), in which the velocity is equal to the average between axial and outside flow velocity, crosses the centers of the puffs;
– in a free submerged jet the puffs submerge through line (plane, surface) of maximum velocity and order in chessboard pattern;
– in a submerged jet laid on solid surface, the puffs also submerge through line (plane, surface) of maximum velocity so the wall boundary layer, actually without any large-scale vorticity, is thinner than the distance from the line (plane, surface) of maximum velocity to the solid surface;
– in interpuff layers (between puffs) near to free boundary(ies), undisturbed flow velocity is kept by inertia without changing of velocity component in the flow direction. If there is no flow belonging to the free boundary(ies), the x-velocity in the layers remains zero. In internal parts of the interpuff layers (belong to centerline), a low-degree polynomial provides the enough precision accepting "smooth-velocity profile" conditions;
– to accept the previous assumption, the puffs may consume all liquid (gas) that it covers during movement. If there is a flow belonging to a free boundary, we need to imaginarily move the initial puff position with the flow velocity. The puff may consume part outside the imaginary contour covered by the actual puff position. The consumed liquid (gas) is used for expansion of the puff, and the interpuff layer belongs to the puff. If the flow runs in contrary flow, the consumed liquid (gas) is also used to fill the "emptiness" that may be formed by contrary flow movement (the considered flow may loose some liquid [gas] to supply the contrary flow);
– a jet in contrary flow may contain at least a reasonable number of puffs (4 or 5) until it is completely destroyed. The jet is fully destroyed near to a puff that stops because of contrary flow influence;
– in a submerged jet laid on solid surface, we can use Israel Shepelev's assumption (Shepelev 1978) that we can eliminate from the consideration of the wall boundary layer. But to obtain the acceptable precision, we need to improve the assumption: We may not cancel the wall boundary layer (original assumption with more than 20% error) but conventionally enlarge the puffs up to the solid surface. The wall boundary layer can be assumed as a vortex film. We can assume the film as ideal lubricant because the "hydraulic friction" has insignificant influence on the flow scale;

– instead of integration by the time, we can integrate along the flow direction by the macro-structure "half-period" – half of puff (in most cases) – or "period" – puff. The precision is acceptable.

Using the assumption above, we solved the following tasks. We theoretically found the basic free jet constant—the tangent of expansion angle $\tan(\beta) = 0.22$. The jet expansion, velocity, and temperature difference profiles and fading were found for free, laid jets, and jets in accompanying or contrary flows. The position of jet separation from a convex surface was estimated. It is proved that jets laid on convex surfaces have less velocity pulsations than jets laid on flat or concave surfaces. The existence conditions of the jet in a contrary flow are specified. The velocity and temperature difference profiles and also heat-transfer coefficient between flows was calculated. All of the results coincide with known experimental data. The simulation results of backflow effect in water heating radiators (causing overheating of rooms), using the last result, provided formulae that included in building norm (Babicheva, Voinalovych, Galinskyi, Gutnichenko, Ivanenko, Maksimov 2014) for thermal modernization of building. This norm specifies requirements for energy efficiency rising of old buildings to decrease heat and chemical (during heat generation from fuel) environmental pollution.

3 EXAMPLE: A JET IN A FLOW

One of the successful examples that can show the principles of the simulation is a non-constrained jet in an accompanying or contrary flow (Gumen, Dovhaliuk & Mileikovskyi 2015). The jet may be assumed as round puffs in chessboard order, as shown in Figure 1.

Let us select puff 1. During a time period $d\tau$, it moves to new position $1'$. If the contour 1 moves with the velocity $u_\perp$ of the ambient flow, it moves to new imaginary position $1''$. The new corresponding positions of points of the puff 1 (A_1 and B_1 – the ends of the diameter normal to the axial velocity; A_{1i} – the points of touching of puffs 1 and i, the center O_1, etc.) are signed with the corresponding primes. The consumed area outside the contour $1''$ but inside the contour $1'$ (grayed on the Figure 1) dA_c must be equal to the total area A_Σ of the puff 1 and the interpuff layer that consists of curvilinear triangles $B_1A_{13}B_3$ and $A_{12}A_{13}A_{23}$ (B_i is the end on the free boundary of the diameter of ith puff normal to the jet axis). If the ambient flow is contrary, the jet also must loose its liquid (gas) to supply the contrary flow. If we move the line B_1B_3 with the contrary flow during the time $d\tau$ the final position of the line will be N_1N_3. So the jet may loose the area dA_m of $N_1N_3B_3B_1$. Finally, we make up the balance $dA_c = dA_\Sigma + dA_m$, where dA_Σ is a growth of the area A_Σ during the time $d\tau$. We found all components by the geometric analysis of Figure 1 in the work (Gumen, Dovhaliuk & Mileikovskyi 2015) and replaced A_Σ and dA_m with the areas of the trapezoid $PQRS$ around the puff 1 and the area of the parallelogram $IJQP$ obtained by the movement of the line PQ with the velocity $u_\perp$ using precisely found proportion coefficients.

The result of expansion angle β tangent $\Theta = \tan(\beta)$ has very good coincidence with the theory of Henrih Abramovich (Abramovich 2011) only for the accompanying or absent ambient

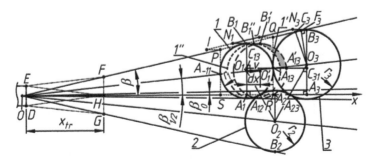

Figure 1.   Chart of a jet in accompanying or contrary flow.

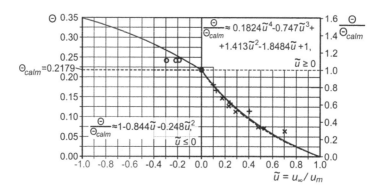

Figure 2. The tangent of the jet expansion Θ, absolute and relative to the tangent Θ_{calm} in the calm environment without a flow dependent on the relative velocity u_∞/u_m: thick solid line—the calculation data, thick dashed line—the theory of Abramovich (2011); signs—experimental data (Abramovich 2011) of $+$ – O. Yakovlevskyi; $\times$ – B. Zhestkov, and other; O – H. Abramovich & F. Vafin.

flow as it is shown in the right half of Figure 2 and contradict with it for the contrary flow. But it has good coincidence with experimental data of different authors for all of the cases. It is interesting that the experimental data on contrary flow are almost nearer to the calculation results of this work than to the theory of Abramovich (2011). Also, the interesting fact is the absence of the experimental data below $u_\infty/u_m = -0.3$.

Using the momentum equation for the inlet opening and the end of the jet initial section *FG* (with constant velocity jet core *EHD*) and the geometrical analysis of puff macrostructure, we found that below $u_\infty / u_m \approx - 0.4$ the jet can destroy before forming.

In this work, we don't repeat the calculation performed in Gumen, Dovhaliuk & Mileikovskyi (2015) because the access to the journal "*BoZPE*" is free (http://www.bud.pcz. czest.pl/bozpe-nr-11520152). But we will pay attention on experimental validation of the obtained result.

Effective way of jet existence test is visual research. We need two ventilation apparatus. The first one is an aerodynamic stand with small nozzle for jet generation with different velocities. The second is a wind machine that generates high diameter even flow to avoid the influence of a jet boundary layer at the bounds. Also, it is important to accurately set both machines coaxially and accurately acquire the jet image with length measurement possibility. It is perfectly to obtain the axial jet section with the jet core image in correlation with macrostructure image around it. The separation of the views obtained at the same time can be performed by color.

4 EXPERIMENTAL TEST OF JET EXISTENCE IN CONTRARY FLOW

For experimental validation, we used the stand in Figures 3 and 4. Aerodynamic stand works in the following way. Air enters it by the smoke collecting funnel 1, passes through gas valve 2 to gasmeter 3 to measure the flow and runs through fan 4. The fan causes flow swirl that cancels by the honeycomb 5. The stabilized flow passes the pressure camera 6 with the grid 7 and microgrid 8 and enters the collector 9 for final stabilization. The stabilization duct 10 is required for the full development of the flow. After it, there is the extension tube 11 to avoid the influence of the pressure camera 9 butt-end. The extension tube 11 is finished by the jet inlet opening 12. For the visualization, there is the smoke machine 13 with nozzle submerged to the smoke collecting funnel 1. For the fan 4 control, there is the aerodynamic stand autotransformer 14 with the softstarter 15. The "ready" dry contact of the softstarter 15 is connected to one of the serially connected outlets of the extender 16. When the softstarter is ready, the outlet would be short circuited and the "Ready" lamp 17 connected to the other outlet of the extender 16 will signal that the aerodynamic stand is ready for experimenting.

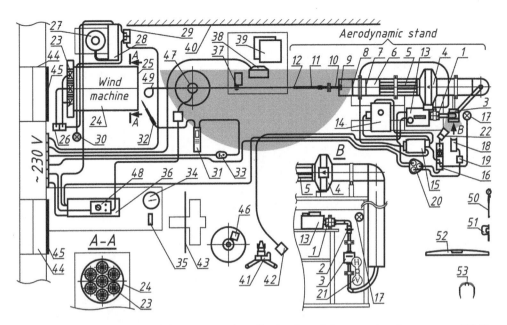

Figure 3. Experimental stand chart: 1 – smoke collecting funnel; 2 – gas valve; 3 – gasmeter (GMS-Arsenal G6 Q_{max} = 10 m³/h, Q_{nom} = 6 m³/h, Q_{min} = 0.08 m³/h, main uncertainty is 3% at $Q \geq 0.5$ m³/h, $t = -20 \cdots +50°C$); 4 – fan (Vent 150 L—direct flow radial fan Ø150 with increased pressure); 5 – honeycomb; 6 – pressure camera (Ø160 × 4); 7 – grid; 8 – microgrid (tulle, less than 0.25 mm cells); 9 – collector (Bernoulli lemniscate shape); 10 – stabilization duct (Dn15, 10 inside diameters length); 11 – extension tube; 12 – jet inlet opening; 13 – smoke machine (American DJ MiniFog with smoke juice EuroLite "P"); 14 – aerodynamic stand autotransformer; 15 – softstarter; 16 – aerodynamic stand outlet block with serially connected outlets; 17 – aerodynamic stand "Ready" signal lamp; 18 – camcorder (Canon Legria HFR306 HD); 19 – camcorder power supply module; 20 – aerodynamic stand outlet block; 21 – letter signal card set; 22 – spotlight with a switch for gasmeter illumination (with 2 W LED lamp to avoid flow meter heating); 23 – seven axial fans (VN-2); 24 – mixing camera (Ø430); 25 – wind machine output with microgrid (tulle, less than 0.25 mm cells); 26 – outlet block of wind machine; 27 – wind machine autotransformer; 28 – softstarter; 29 – outlet block of wind machine with serially connected outlets; 30 – wind machine "Ready" signal lamp; 31 – thermoelectroanemometer (datalogger Testo 445); 32 – thermoelectroanemometer probe (Testo 0635 1049 "Hot Bulb" $v = 0,...,10$ m/s, uncertainty ± (0.03 m/s ± 5% m.v.); $t = -20,..., + 70°C$, uncertainty ± 0.3°C); 33 – thermoelectroanemometer power supply module (Energiia EH723, 12 V, 800 mA, linear voltage stabilizer); 34 – barometer (BAMM-1, $p_b = 80,..., 106$ hPa, scale factor 0.1 hPa, uncertainty basic – ± 0.2 hPa—complementary – ±0.5 hPa); 35 – thermohygrometer (TESTO-608 H1, $\varphi = 10,..., 95\%$, absolute uncertainty ± 3%, $t = 0,..., + 50°C$, uncertainty ± 0.5°C); 36 – dual-conversion uninterruptible power supply (MGE UPS system Pulsar 1500); 37 – red laser pointer with flat ray optical system (SDLaser 303, 250 mW, 60° aperture angle, class 3 laser product); 38 – green (20 W) LED spotlight; 39 – green ray filters to exactly balance red laser and green spotlight; 40 – black screen; 41 – camera (Canon EF 50 mm 1:1.8 lens with appropriate lens hood); 42 – foot switch for smoke machine 13 remote control; 43 – experiment label stand; 44 – windows for smoke elimination between experiments; 45 – laser ray protective shields; 46 – laser self-setting level (Bosch Quigo, ±0.8 mm/m or 2′45″, class 2 laser product); 47 – vertical flat laser on the floor (250 mW flat laser module with 180° aperture angle); 48 – laser module voltage control and dimmer (PIKO ME005, 1.2 A, 1.7,..., 12 V); 49 – measuring tape (Kondor 8 m × 25 mm); 50 – hole-gauge (KI, 10–18 mm, GOST 868–63, 0.01 mm); 51 – micrometer (LIZ, 0,..., 25 mm 0.002 mm); 52 – level (masonry level Kapro 24″ 60 cm); 53 – cyan laser safety glasses.

For the flow meter 18 readings recording in the time, there is the camcorder 18 with its power supply 19 connected to the aerodynamic stand outlet block 20 that also supplies power to the smoke machine 14 and the extender 16. To record the experimental number with the flow meter 3 readings, there is the letter signal card set 21. For the meter 3 display and signal card 21 illumination, there is the low power (to avoid the meter 3 heating) spotlight 22.

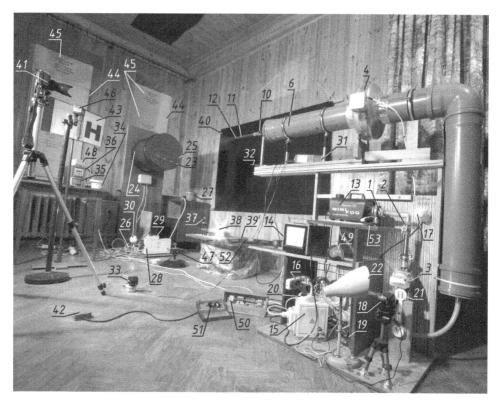

Figure 4. Experimental stand. Numerical labels are equal to Figure 3.

The wind machine contains seven axial fans 23. In the mixing camera 24, the swirled flows from the fans 23 are mixed, and finally the swirling is canceled by the microgrid 25. The velocity control (26–30) acts similarly to the aerodynamic stand control. In the inlet opening 12 of the aerodynamic stand, we can measure axial velocity by the thermoelectroanemometer 31 with the hot bulb probe 32. The thermoelectroanemometer 31 is powered by the power supply 33 with linear voltage stabilizer to avoid power noises influence on the measuring results.

To correct the velocity readings, there is the barometer 34, and for the ambient conditions measurement, there is the thermohygrometer 35. To avoid velocity drift in the aerodynamic stand and wind machine, because of voltage change, there is a dual-conversion uninterruptible power supply unit 36 that supplies power only to the fans through fan controls.

For the visualization, there is the jet illumination block with the red flat laser 37 to obtain red jet section and the green LED spotlight 38 to illuminate the whole jet. There is the camera with video recording function 41 on a tripod to obtain jet pictures (Figs. 5–8) and videos. Near camera 41, there is the foot switch 42 to control smoke release remotely from the smoke machine 13. To correspond the shots and flow meter 3 reading, there is the experimental label stand 43 with changeable letters. The shot of the corresponding letter may be obtained before an experiment starts.

To eliminate smoke between experiments, laser 37 may be deactivated and only after that the air-tight windows 44 with the laser ray protective shields 45 may be opened until the smoke will run out, and the air will be enough clear to start the next experiment. For safety reasons, laser 37 will be activated only after closing windows 44. For the aerodynamic stand and the wind machine positioning, there are horizontal laser safe-setting level 46 and the vertical flat laser 47 with the voltage control (dimmer) 48 to avoid blinding.

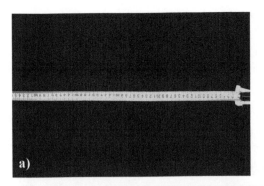

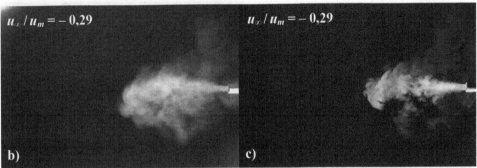

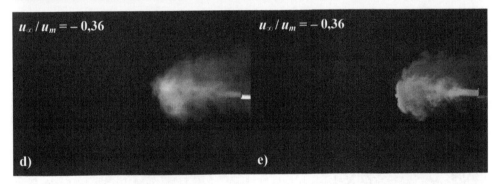

Figure 5.   Jet shots: a – basic shot; b & c – accordingly general view and section at $u_\infty / u_m = 0.29$; d & e – accordingly general view and section at $u_\infty / u_m = 0.36$.

To focus the camera 41 and obtain the linear characteristic of shots, the measuring tape 49 may be put between inlet centers of the aerodynamic stand and the wind machine. A camera 41 lens may be autofocused on tape 49, switched to the manual focus mode, and the basic shot (Fig. 5a) may be obtained. As the camera will not be moved, the jet pictures may be compared with the basic shot to obtain the jet length (range).

For hole-12-diameter measurement, there are hole-gauge 50 and micrometer 51. Also we use precise level 52 for fine-setting of the stands. To avoid blinding by the red laser rays from the positions 37 and 46, there are cyan laser safety glasses 53.

As is shown on Figure 5 at $u_\infty/u_m = -0.36$, there is a formed short-range jet. In Fig. 6, at $u_\infty/u_m = -0.42$, the jet is unstable with periodic change of range. The measurement of averaged velocity may require an unreasonable number of repetitions because of the very high level of pulsations. Figure 7 shows that at $u_\infty/u_m = -0.53$; the instability becomes deeper until periodic full destroys—development cycles. And at $u_\infty/u_m = -0.61$, the jet cannot develop as it is shown in Figure 8. So, the calculation result (stable jet existence until $u_\infty/u_m \approx -0.4$) is validated.

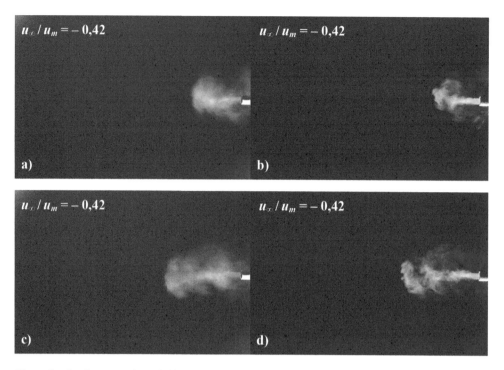

Figure 6.   Jet shots at $u_\infty / u_m = 0.42$: a & b – accordingly general view and section for one time moment; c & d – the same at different time moment: basic shot is shown in Figure 5a.

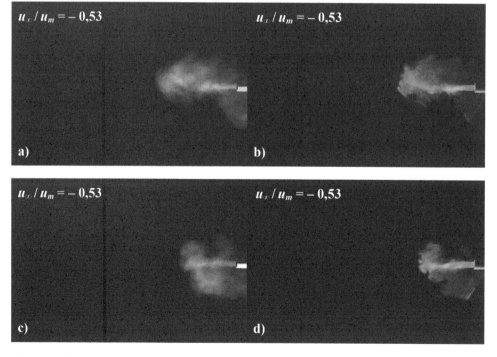

Figure 7.   (*Continued*)

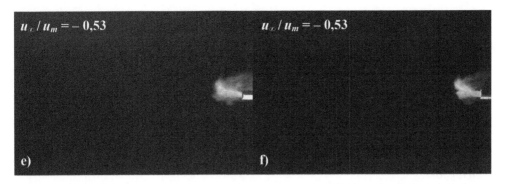

Figure 7. Jet shots at $u_\infty / u_m = 0.53$: a & b – accordingly general view and section for one time moment; c & d – the same at different time moment; e & f – the same at another time moment: basic shot is at Figure 5a.

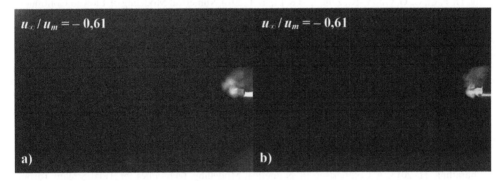

Figure 8. Jet shots at $u_\infty / u_m = 0.61$: a & b – accordingly general view and section: basic shot is at Figure 5a.

5 CONCLUSIONS

An approach for simplified simulation of flows with large-scale vorticity is proposed on the basis of geometric and kinematic analysis of the flow macrostructure. It simulates subsonic jets and boundary layer between flows. The assumptions list, given in this work, provides results that coincide with experimental data.

The jet simulation in accompanying or contrary flow is an example of the approach use. The predicted jet destruction near to $u_\infty/u_m \approx -0.4$ is experimentally validated by visual researches.

REFERENCES

Abramovich, H.N. 2011 *Teoriia turbulentnykh strui [Theory of Turbulent Jets]*. Minsk: Ekolit.
Alaimo, A., Esposito, A., Messineo A., Orlando, C. & Tumino, D. 2015. 3D CFD Analysis of a Vertical Axis Wind Turbine. *Energies* 8(4): 3013–3033.
Ali, A.A.A.A., Megri, A.C., Dellenback, P.A. & Yu,Y. 2014. Use of Computational Fluid Dynamic to Predict Airflow and Temperature Distribution in a Residential Building With an Under Floor Air Distribution System. *American Journal of Engineering and Applied Sciences* 7(1): 171–184.
Anggiansyah, R. & Prabowo 2014. Studi Numerik Pengaruh Posisi Sudut Obstacle Berbentuk Rectangular terhadap Perpindahan Panas pada Tube Banks Staggered. *Jurnal Teknik Pomits* Vol. 3. No. 2: B-186 – B–191.
Arifin, M., Pasec, A.D. & Eddy, Z. 2015. Geometry Analysis and Effect of Turbulence Model on the Radial Rotor Turbo-Expander Design for Small Organic Rankine Cycle System. *Journal of Mechatronics, Electrical Power, and Vehicular Technology* 6(1): 39–48.

Babicheva, P., Voinalovych, I., Galinskyi, O., Gutnichenko, T., Ivanenko, V. & Maksimov, A. 2014 DSTU-N B V.3.2–3:2014. Nastanova z vykonannia termomodernizatsii zhytlovykh bydynkiv [Direction for Performing of Thermal Modernization of Residential Buildings]. Kyiv: Ukrarkhbudinform.

Bahirat, S. & Joshi, P.V. 2014. CFD Analysis of Plate Fin Tube Heat Exchanger for Various Fin Inclinations. *International Journal of Engineering Research and Applications*. Vol. 4 Iss. 8 (Version 3), August 2014. 116–125.

Bose, S.K. 2015. Shallow Water Turbulent Surface Wave Striking an Adverse Slope. *Acta Geophysica* 63(4): 1090–1102.

Cheviron, B. & Moussa, R. 2015. Determinants of modelling choices for 1-D free-surface flow and erosion issues in hydrology: a review. *Hydrology and Earth System Sciences Discussions* 12(9): 9091–9155.

Gumen, O., Dovhaliuk, V. & Mileikovskyi, V. 2015 Geometric Modelling of Turbulent Flat Jets in Accompanying and Contrary Flows. *Budownictwo o zoptymalizowanym potencjale energetycznym* 1(15): 70–77.

Ismail, K.A.R. & Batalha, T.P. 2015. A comparative study on river hydrokinetic turbines blade profiles. *International Journal of Engineering Research and Applications* 5(5): 01–10.

Jianxi, Y. 2015. Investigation on hydrodynamic performance of a marine propeller in oblique flow by RANS computations. *International Journal of Naval Architecture and Ocean Engineering* 7(1): 56–69

Jing, H.-F, Li, C.-G, GuoY.-K., Zhu, L.-J & Li, Y.-T. 2014 Numerical Modeling of Flow in Continuous Bends from Daliushu to Shapotou in Yellow River. *Water Science and Engineering* 7(2): 194–207.

Kiczko, A., Kubrak, J., & Kubrak, E. 2015. Experimental and numerical investigation of non-submerged flow under a sluice gate. *Annals of Warsaw University of Life Sciences—SGGW* 47(3): 187–201.

Kwang-Jun, P., Seunghyun, H., Jaekwon, J., Taegu, L., Yeong-Yeon, L., Haeseong, A. & Suak-Ho, V. 2015. Investigation on the wake evolution of contra-rotating propeller using RANS computation and SPIV measurement. *International Journal of Naval Architecture and Ocean Engineering* 7(3): 595–609.

Lee, S. 2015 A numerical study on ship-ship interaction in shallow and restricted waterway. *International Journal of Naval Architecture and Ocean Engineering* 7(5): 920–938.

Li, D.Y., Gong, R.Z., Wang, H.J., Zhang, J., Wei, X.Z. &. Shu, L.F. 2016. Numerical Investigation in the Vaned Distributor under Different Guide Vanes Openings of a Pump Turbine in Pump Mode. *Journal of Applied Fluid Mechanics* 9(1): 253–266.

Masters, I., Williams, A., Croft, T.N., Togneri M., Edmunds, M., Zangiabadi, E., Fairley, I., & Karunarathna, H. 2015. A Comparison of Numerical Modelling Techniques for Tidal Stream Turbine Analysis. *Energies* 8(8): 7833–7853.

Nakayama, H., Takemi, T. & Nagai, H. 2015. Large-eddy simulation of turbulent winds during the Fukushima Daiichi Nuclear Power Plant accident by coupling with a meso-scale meteorological simulation model. *Advances in Science and Research; Open Access Proceedings; 14th EMS Annual Meeting & 10th European Conference on Applied Climatology (ECAC), Praha, 06–10 October 2014* 12: 127–133.

Shepelev, I.A. 1978. *Aerodynamika vozdushnyh potokov v pomeshchenii [Aerodynamics of Air Flows in a Room]*. Moscow: Stroiizdat.

Smagorinsky, J. 1963. General Circulation Experiments with the Primitive Equations. *Monthly Weather Review* 91(3): 99–164.

Stepanov, B.L., Pešenjanski, I.K. & Spasojević M.D. 2015. Scandinavian Baffle Boiler Design Revisited. *Thermal Science* 19(1): 305–316.

Sungwook, L. & Booki, K. 2015. A numerical study on manoeuvrability of wind turbine installation vessel using OpenFOAM. *International Journal of Naval Architecture and Ocean Engineering* 7(3): 466–477.

Tian, W., Song, B., VanZwieten, J.H. & Pyakurel, P. 2015. Computational Fluid Dynamics Prediction of a Modified Savonius Wind Turbine with Novel Blade Shapes. *Energies* 8(8): 7915–7929.

TienPhuc, D., ZhengQi, G. & Zhen, C. 2016. Numerical Simulation of the Flow Field around Generic Formula One. *Journal of Applied Fluid Mechanics*, Vol. 9, No. 1: 443–450.

Tkachuk, A.Y. 2001. Rozrakhunkova model userednenogo rukhu v turbulentnii zoni ploskykh i visesymetrychnykh prysinnykh prymezhovykh shariv [Computational model of averaged flow in turbulent zone of flat and axisymmetric wall boundary layers]. *Ventyliatsiia, osvitlennia ta teplogazopostachannia.* Iss. 2.

Toja-Silva, F., Peralta, C., Lopez-Garcia, J., Navarro, J. & Cruz, I. 2015. On Roof Geometry for Urban Wind Energy Exploitation in High-Rise Buildings *Computation (Basel)* 3(2): 299–325.

Tongpun, P., Bumrungthaichaichan, E. & Wattananusorn, S. 2014. Investigation of entrance length in circular and noncircular conduits by computational fluid dynamics simulation. *Songklanakarin Journal of Science and Technology* 36(4): 471–475.

Warzecha, M., Merder, T. & Warzecha, P. 2015. Investigation of the Flow Structure in the Tundish with the Use of RANS and LES Methods. *Archives of Metallurgy and Materials* 60(1): 215–220.

Hydraulic Engineering IV – Xie (Ed.)
© *2016 Taylor & Francis Group, London, ISBN 978-1-138-02948-4*

Author index